Queueing Theory for Telecommunications

Attahiru Sule Alfa

Queueing Theory for Telecommunications

Discrete Time Modelling of a Single Node System

Attahiru Sule Alfa
Dept. of Electrical & Computer Engineering
University of Manitoba
Winnipeg, Manitoba
Canada R3T 5V6
alfa@ee.umanitoba.ca

ISBN 978-1-4899-8744-0 ISBN 978-1-4419-7314-6 (eBook)
DOI 10.1007/978-1-4419-7314-6
Springer New York Dordrecht Heidelberg London

Printed on acid-free paper

Springer is part of Springer Science+Business Media (www.springer.com)

This is dedicated to all my parents, my wife,
my children and my grandchildren.

Preface

It is not as if there are not many books already on queueing theory! So, why this one – an additional book? Well, generally queueing models and their analyses can be presented in several different ways, ranging from the highly theoretical base to the very applied base. I have noticed that most queueing books are written based on the former approach, with mathematical rigour and as a result sometimes ignoring the application oriented reader who simply wants to apply the results to a real life problem. In other words, that reader who has a practical queueing problem in mind and needs an approach to use for modelling such a system and obtain numerical measures to assist in understanding the system behaviour. Other books that take the purely applied approach usually oversimplify queueing models and analyses to the extend that a reader is not able to find results useful enough for the major applications intended. For example, the models are limited to simple unrealistic arrival and service processes and rules of operation, in an attempt to obtain very simple analysis which may not capture the key behaviour of the system. One of the challenges is to present queueing models and clearly guide the reader on how to use available mathematical tools to help in analyzing the problems associated with the models. To this end the contents of this book are mainly presented in a form that makes their applications very easy to see, as such it is written to make queueing analysis easier to follow and also to make the new developments and results in the field more accessible to pragmatic user of the models.

Nearly every queueing model can be set up as a Markov chain, even though some of them may end with huge state space. Markov chains have been very well studied in the field of applied stochastic processes. There are a class of Markov chains for which the matrix-analytic methods (MAM) are very suitable to their analysis. Most queueing models fall in that category. Hence there is definitely value in developing queueing models based on Markov chains and then using MAM for their analyses. The philosophy behind this book is that for most practical queueing problems that can be reduced to single node system, we can set them up as Markov chains. If so, then we can apply the results from the rich literature on Markov chains and MAM. One whole chapter is devoted to Markov chains in this book. That chapter focuses on Markov chain results and those specific to MAM that will be needed for the single node analysis. So most of the analysis techniques employed in this book are based on MAM. Once a while the z-transform equivalent of some of the results are developed as alternatives. However, the use of transforms is not a major focus of this book. The book by Bruneel and Kim [28] focused on transform approach and the reader who has interest in that topic are referred to that book. Additional results on discrete time approach for queues using z-transforms can be found by visiting Professor Bruneel Herwig's website.

It is well known that single node queues do give some very good insights to even the very complex non-single node queues if properly approximated. So single node queues are very important in the field of queueing theory and are encountered very frequently. This book focuses on single node queues for that reason. Queueing networks in discrete time are dealt with in the book by Daduna [34].

One important area where queueing theory is applied very frequently is in the field of telecommunications. Telecommunication systems are analyzed in discrete time these days because it is based mainly on discrete technology; time is slotted and the system has shifted from analogue technology to discrete one. Hence discrete time queueing models need special consideration in the fields of queueing and telecommunications.

The first chapter of this book introduces single node queueing systems and discusses discrete time briefly. We believe, and also want to show in this queueing book, that most single node queues in discrete time can be set up as discrete time Markov chains, and all we need to do is apply the results from Markov chain. As a result we present, in Chapter 2, the key results from discrete time Markov chains, especially those related to MAM, that will be used to analyze our single node queueing systems. Generally, for most practical queueing problems we find ourselves turning to recursive schemes and hence needing to use some special computational procedures. Emphasis is placed on computational approaches for analysing some classes of Markov chains in Chapter 2. As we get into Chapters 3 and 4 where single node queues are dealt with the reader will see that the real creative work in developing the queuing models is in setting them up as Markov chains. After that we simply apply the results presented in Chapter 2. This is what is special about this book – that all we need is to understand some key results in discrete time Markov chains and we can analyze most single node queues. To summarize the philosophy used in this book, Chapter 2 actually contains the tools needed to analyze single node queues considered in this book, and the materials presented in Chapters 4 and 5 are mainly developing and also showing how to develop some key single server queueing models. For each of the models presented in Chapters 4 and 5, the reader is pointed to the Markov chain results in Chapter 2 that can be used for the queueing models' analyses. Chapter 3 presents general material that characterize queueing systems, including the very common distributions used to characterize the interarrival and service times.

Winnipeg, Manitoba, Canada

Attahiru S. Alfa
May 2010

Acknowledgements

First I wish to thank all the people that have contributed to my learning of queueing theory, people like Do Le Minh, W. R. Blunden and Marcel Neuts, and several colleagues such as Srinivas Chakravarthy, V. Ramaswami, Qi-Ming He, and the list goes on. My sincere thanks goes to several of my graduate students over the years – they made me continue to maintain my keen interest in queueing theory. I sincerely thank Telex Magloire Ngatched Nkouatchah and Haitham Abu Ghazaleh for helping with going through the final stage draft of this book. Professor Bong Dae Choi of Korea University, Professors Vicente Casares Giner and Vicent Pla, both of the Universidad Politécnica de Valencia were kind enough to let me spend portions of my sabbatical leave of 2009-2010 in their respective universities with fundings and facilities to support my visit, during a period that I was trying to complete this book. To them I am most grateful because the timing was perfect.

Acknowledgments

Contents

Chapter 1
Introduction

1.1 Introduction

A queueing system is a system in which items come to be processed. The words "item" and "processing" are generic terms. An "item" could refer to customers. They could be customers arriving at a system such as a bank to receive service (be "processed"). Another example is in a manufacturing setting where an item could be a partially completed part that needs to be machined and is thus sent to a station where it is duly processed when possible. Most of us go through city streets driving vehicles or as passengers in a vehicle. A vehicle going through a signalized intersection is an item that needs to be served and the service is provided by the intersection in the form of giving it a green light to go through. The type of queueing systems of interest to us in this book are the types encountered in a communication system. In such a system, the item could be a packet arriving at a router or switch and needs its headers read so it can be forwarded to the appropriate next router or final destination. In this introductory chapter we will use all the similies of queueing system characteristcs interchangeably so as to emphasize that we are dealing with a phenomenon that occurs in diverse walks of life. Generally in a queueing system there is a physical or a virtual location, sometimes moving, where items arrive to be processed. If the processor is available the item may be processed immediately otherwise it has to wait if the processor is busy. Somehow, the waiting component that may occur if the processor is busy seems to dominate in the descriptor of the system, i.e. "queueing" or "waiting". In reality, this system can also be called a service system – sometimes a better descriptor. But we will stay with queueing system to maintain tradition.

Nearly all of us experience queueing systems directly or indirectly; directly through waiting in line ourselves or indirectly through some of our items waiting in line, such as a print job waiting in the printer buffer queue, or a packet belonging to a person waiting at a router node for processing. In any case, we all want our delays to be minimum and also not be turned away from the system as a result of the buffer space not being available. There are other types of factors that concern

A.S. Alfa, *Queueing Theory for Telecommunications*,
DOI 10.1007/978-1-4419-7314-6_1, © Springer Science+Business Media, LLC 2010

most other people, but these two are the major ones. Of course we all know that if those managing the system can provide "sufficient" resources to process the items then the delays or denial of service possibilities will be small. However, the system manager that provides the service is limited to how much resources to provide since they do cost money. Understanding how queueing systems behave under different scenarios is the key reason why we model queueing systems. This book intends to continue in that vein, but it focusses mainly on what we call a single node queueing system.

1.2 A single node queue

A single node queue is a queueing system in which a packet arrives to be processed in one location, such as at a router. After the packet has been processed it does not proceed to another location for further processing, or rather we are not interested in what happens to it at other locations after the service in the current location is completed. However the packet may re-enter the same location for processing immediately after its processing is completed. If it proceeds to another location for another processing and later returns to the first location for processing then we consider this return as a new arrival and the two requests are not directly connected. So a single node system has only one service location and service is provided by the same set of servers in that location even if a packet re-enters for another service immediately after finishing one. We give examples of single node queues in Figures 1.1, 1.2, 1.3 and 1.4. In Figure 1.1 we have the top portion showing a simple single server queue, where the big circle represents the server, the arrow leading to the server shows arrivals of items and the arrow leaving the server shows departure of items that have completed receiving service.

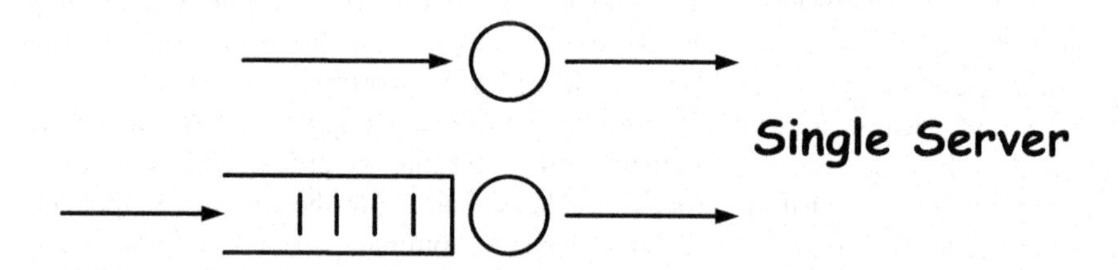

Fig. 1.1 A Single Node, Single Server Queue

Figure 1.1 (the lower part) is an example of a simple single node queue. There is only one server (represented by the circle) at this node. Packets arrive into the buffer (represented by the open rectangle) and receive service when it is their turn. When they finish receiving service they may leave. A queueing system that can be represented in this form is a bank with drive-in window where there is only one ATM machine to be used and vehicles queue up to use it.

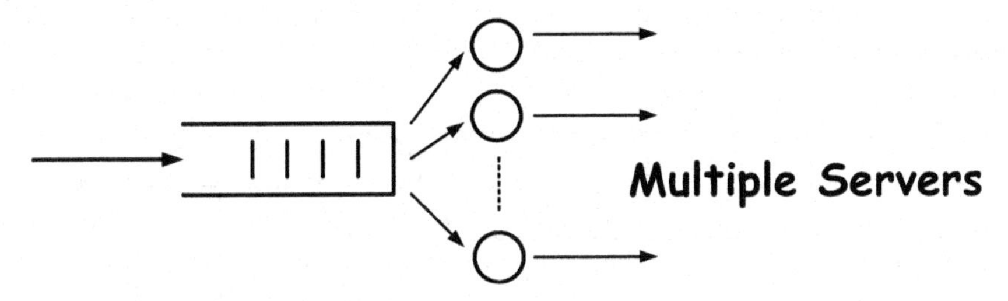

Fig. 1.2 A Single Node, Multiple Parallel Servers Queue

In Figure 1.2 we have a single node queue with several servers in parallel. A packet may be sent to any of the servers for processing. This type of queueing system is very common at the in-buffer queue of a switch in communication networks, the check-out counter in a grocery store, at a bank where tellers attend to customers, a commercial parking lot where the parking bays are the servers and the service times are durations of parking, etc.

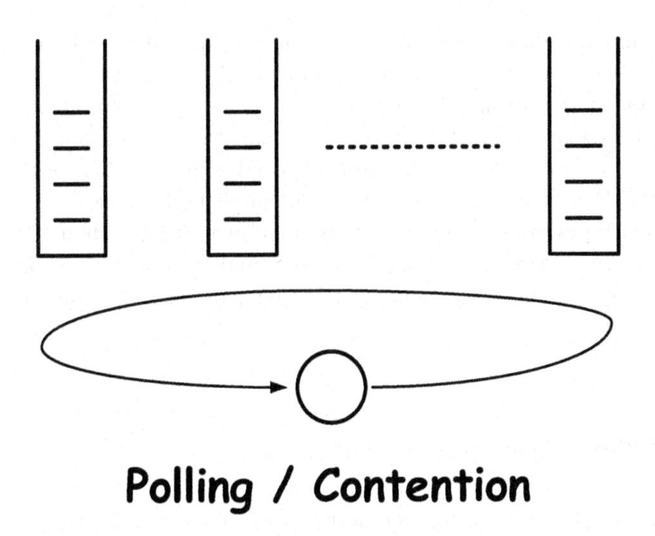

Fig. 1.3 A Single Node, Multiple Parallel Queues and One Server

Figure 1.3 is a single node queue in which there are several buffers for different queues and a server attends to each queue based on a pre-scheduled rule, e.g. round-robin, table polling, etc. This is called a polling system. This type of queueing system is very common in telecommunication systems in the medium access control (MAC) where packets from different sources arrive at a router and need to be processed and given access to the medium. The server (router) attends to each source according to a pre-scheduled rule. Another example of this is a traffic intersection with lights – each arm of the intersection is a queue of its own and the server, which is the light, gives green light according to a pre-scheduled rule to each arm. The example diagram of Figure 1.3 shows only one server, but we could have more than one server providing the same service.

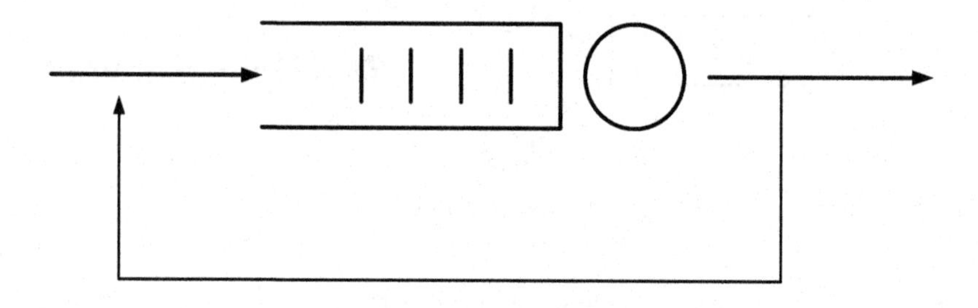

Fig. 1.4 A Single Node, One Server and Feedback

Finally Figure 1.4 is a single node queue in which an item, after completing a service, may return to the same system immediately for another service. This is a feedback queue. For example consider a manufacturing system in which an item after going through a manufacturing process is inspected instantaneously and placed back in the same manufacturing queue if it is deemed to be defective and needs to be re-worked. These are just a few examples of single node queues.

What other types of queueing system exist and what makes them different from single node queues? Single node queues are the smallest units that can be classified as a queueing system. Other types of queueing systems are configurations of several single node queues.

1.3 A tandem queueing systems

Consider several single node queues in series, say N of them labelled $Q_1, Q_2, \cdots, Q_N$. Items that arrive at Q_1 get processed and once their service is completed they may return to Q_1 for another service, leave the system completely or proceed to Q_2 for an additional service. Arriving at Q_2 are items from Q_1 that need processing in Q_2 and some new items that join Q_2 from an external source. All these items arriving in Q_2 get processed at Q_2. Arriving items go through the N queues in series according to their service requirements. We ususally reserve the term tandem queues for queues in which items flow in one direction only, i.e. $Q_1 \rightarrow Q_2 \rightarrow \cdots \rightarrow Q_N$, i.e. we say $p_{i,j}$ is the probability that an item that completes service at node Q_i proceeds to node Q_j, with $p_{i,j} \geq 0,\ j = i, i+1$ and $p_{i,j} = 0,\ j < i,\ j < i+1$, keeping in mind that $\sum_{i=1}^{N} p_{i,j} \leq 1$, since an item may leave the queue after service at node Q_i with probability $1 - \sum_{i=1}^{N} p_{i,j} \leq 1$, i.e. $1 - p_{i,i+1} \leq 1$. An example of a tandem queueing system is illustrated in Fig 1.5.

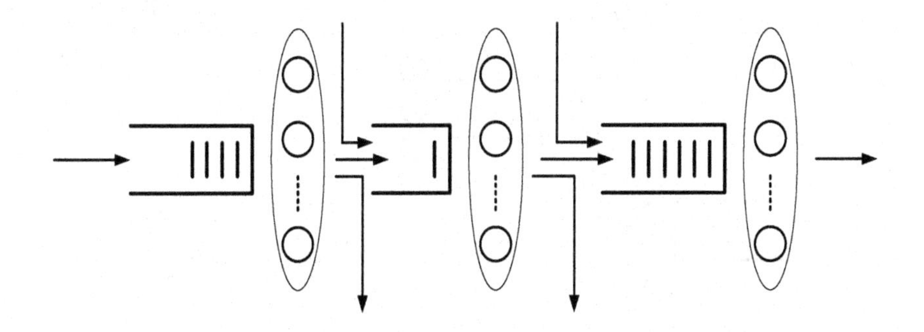

Fig. 1.5 Example of a Tandem Queue

As one can see, a tandem queueing system can physically be decomposed into N single node queues. Even though the analysis as a set of single node queues is involved and may also end up only as an approximation, this approach has been adopted considerably in the literature [87]. For other results on discrete-time analysis of tandem queues, see [86]. A real life example of a tandem queueing system is a virtual circuit in a communication system. Another one is a linear production line in a manufacturing system. One example most of us come accross nearly everyday is that of a path of travel. Once we have decided on our route of travel, say by automoble in a city, then we see this as a linear path and each of the intersections along the path form a set of single node queues and all together form a tandem queueing system.

1.4 A network system of queues

Finally, we have a network system of queues which is a more general class of queueing systems. In such a system we may have N nodes. Items can arrive at any of the N nodes, e.g. node Q_i for processing. After processing at node Q_i the item may leave the system or proceed to another node Q_j for processing or even leave the system entirely. Here we have $p_{i,j} \geq 0,\ \forall (i,j)$, and $1 - \sum_{i=1}^{N} p_{i,j} \leq 1$ is the probability that a job leaves the system at node Q_i. There is no restriction as to which node an item may proceed to after completing service at a particular node, unlike in the case of the tandem queueing systems that has restrictions in its definition. Figure 1.6 illustrates an example of what a network of queues looks like.

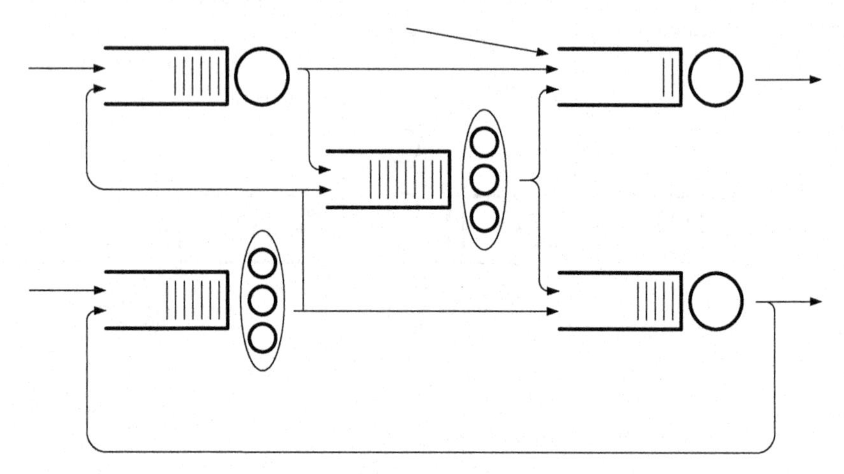

Fig. 1.6 Example of a network of Queues

Most real life queueing systems are usually networks. An example is the communication packet switching system, road networks, etc. Once more one can see that a network system of queues is simply a collection of several single node queues. Sometimes a queueing network is decomposed into several single node queues as an approximation for analysis purposes. This is because analysing queueing networks or even tandem networks could be very involved except in the simpler unrealistic cases. As a result decomposing tandem queues and queueing networks has received a considerable amount of attention in the literature. This is why we decided to focus on single node queues and develop the proper analysis for them in this book. The book by Daduna [34] treats a class of discrete-time queueing networks with geometric interarrival times and geometric services.

1.5 Queues in Communication networks

There are so many types of queueing systems in communication networks. In fact one may say in order to provide an effective and efficient communication system understanding the underlining system that governs the performance of the communication is only second to the development of the equipments needed for the system. In what follows we present just a few examples of communication networks, ranging from the very old to the very new systems.

- **Old POTS:** The plain old telephone system (POTS) is one of the oldest systems that actually established the need for queueing models. The works of the pioneering queueing analyst, Erlang [39], [40], was a result of trying to effectively

understand the switching system of the POTS and how to determine the number of trunks required to effectively and efficiently operate such a system. That is an example of a queueing system at a high level. After switching is carried out at each node, a circuit is established to connect to communicating entities. In the POTS there are several "mini" queueing system within. For example, trying to get a dial tone actually involved a queueing process.
Queueing problems as seen in the POTS had to do with voice traffic – a more homogeneous type of traffic. Only one class of traffic was considered as important or as the main driver of the system.

- **Computer communication networks:** In the computer communication world data was moved in the form of packets. For data packets, instead of circuit switching as in the POTS case, packet switching was employed. Circuits were not established, but packets were given headers and sent through the network via different paths and later assembled at the end (destination) node. Still packets go through a queueing system and depending on the types of packets been sent they may have some priorities set.
- **Integrated Services Digital Networks (ISDN):** In the late 1970s and early 1980s a considerable effort was focussed on trying to integrate both the voice and data networks into one so as to maximize resources. The ISDN project was started and out of it came also the Asynchronous Transfer Mode (ATM), an idea that packetizes both voice and data to small size $(48+5\text{kb})$ cells and sent through the system either through a virtual circuit or as a packet switched data. With ATM came serious considerations for different classes of traffic and priority queues became very important in telecommunication queues. In fact, the advent of ISDN, ATM and the fact that communication systems started to be operated as digitial systems (rather than analogue) led to discrete time systems becoming more prevalent.
- **Wireless networks:** When wireless communication networks became accessible to the paying public the modelling of the queueing systems associated with them opened up a new direction in queueing systems. With the wireless system came mobility of customers. So customers or mobile stations that are receiving service while they are moving could end up switching to a different server (when they change location) without completing their service at the previous server, i.e. mobile stations go through a handoff process.
- **Cognitive radio networks:** Cognitive radio network is currently one of the recent topics of interest in communication systems that opens up the study of a new type of queueing systems. In this system we have several channels - used by primary users and can only be used by secondary users when the channels are free. Here the secondary users keep searching for channels that are available for use (free) at some points in time. So this is a queueing system in which the server is available only intermittently and the customer (secondary user) has to find which channel (server) and when it is available, keeping in mind that other users (secondary users) may be attempting to use the same channel also.

These are just few of the queueing systems encountered in telecommunication systems.

Trying to present material in this book specifically for telecommunication systems would be limiting the capabilities of the analysis techniques in it. Keeping in mind that despite the fast changing world of telecommunications the fundamental queueing tools available seem to keep pace or sometimes are way ahead of the communication technologies, I decided to keep the book mainly as a tool for telecommunication systems analysis rather than a book on telecommunications systems queues. I present the matrial as a fundamental discrete time single node queue analysis. The results can be used for different telecommunication systems as they arise or appropriately modified to suit the analysis of the new systems. As we see from this section, telecommunication systems technologies ranging from the POTS to cognitive radio network all use queueing models and these models are based on the same stochastic models in concept. I hope the results and analysis techniques in this book can be applied for a long time as the telecommuniction system evolves.

1.6 Why are queueing systems of interest?

Queueing systems are of interest mainly to two categories of people; 1) the users of the systems (customers) and 2) the service providers. The customers need (want) to use the system and minimize their delays, loss (or blocking) probabilities, etc. On the other hand the service provider wishes to minimize its cost of providing the service to customers while ensuring that the customers are "reasonably" satisfied. For example, the service providers do not want to assign too many servers that results in a high percentage of them becoming idle most of the time, they do not want to provide an extremely large buffer size which is often unoccupied, etc. – and they do this without knowing exactly the workload to be submitted by customers because that component is stochastic. Sometimes the rule used to oprerate the system is a major factor in achieveing the proper goals. For example, just by changing the operation rule from first-come-first-served to a priority system will change how the system is perceived by the customers. So the need to understand how to operate a queueing system while achieving all these conflicting goals, is necessary and needs a very good understanding of the systems.

1.7 Discrete time vs Continuous time analyses

Most queueing models in the literature before the early 1990s were developed in continuous time. Only the models that were based on the embedded Markov chain studied the systems in discrete times – models such as the M/G/1 and GI/M/1 which were well studied in the 1950s [61] [62]. The discrete time models developed before then were few and far between. The earlier discrete time models were those by Galliher and Wheeler [44], Dafermos and Neuts [35], Neuts and Klimko [83], Minh [78], just to name a few. However, researchers then did not see any major reasons to study queues in discrete time, except when it was felt that by doing so difficult models became easier to analyze. Examples of such cases include the embedded systems and the queues with time-varying parameters.

Modelling of a communication system is an area that uses queueing models signi-ficatly and continuous time models were seen as adequate for the purpose. However now that communication systems are more digital than analogue and we work in time slots discrete modelling has become more appropriate. Hence new results for discrete time analysis and books specializing on it are required.

Time is a continuous quantity. However, for practical measurement related pur-poses, this quantity is sometimes considered as discrete, especially in modern co-munication systems where we work in time slots. We often hear people say a system is observed every minute, every second, every half a second, etc. If for example we observe a system every minute, how do we relate the event occurrence times with those events that occur in between our two consecutive observation time points? This is what makes discrete time systems slightly different from continuous time systems. Let us clarify our understanding of continuous and discrete view of time.

1.7.1 Time

Consider time represented by a real line. Since time is continuous we are not able to observe any system at time t. We can only observe it between time t and $t+\delta t$, where δt is infinitesimally small. In other words, it takes time to observe an event, irrespective of how small this observation time is. This is because by the time we finish observing the system, some additional infinitesimal time δt would have gone by. Hence, in continuous time, we can only observe a system for a time interval, i.e. between times t and time $t+\delta t$. Note that δt could be as small as one wishes.

In discrete time, on the other hand, our thinking is slightly different. We divide time up into finite intervals. These intervals do not have to be of equal lengths and the time points may or may not represent specific events. We can now observe the system at these points in time. For example, we observe the system at time t_1, t_2, t_3, ... , t_n, t_{n+1}, ..., implying that everything that occurs after time t_n and throughtout the interval up to time t_{n+1}, is observable at time t_{n+1}; for example all arrivals to a system that occur between the interval $(t_n, t_{n+1}]$ are observed, and assume to take place, at time epoch t_{n+1}. Another way to look at it is that nothing occurs between the interval $(t_n, t_{n+1}]$ that is not observable at time epoch t_{n+1}. By this reasoning we assume the picture of the system is frozen at these time points $t_1, t_2, \cdots, t_n, \cdots$, and also assume that it takes no time to freeze a picture. In addition, we only know what we observe at the time points and they constitute the only knowledge we have about the system. Whatever activities occur between our last observation of the sys-tem at time t_n, and our next observation at time t_{n+1}, are assumed to occur at time t_{n+1}. We normally require that an item is in the system for at least one unit of time, e.g. a data packet requires at least one time slot for processing, and hence it will be in the system for at least that duration before departing. Hence, our discrete time analysis by definition will be what Hunter [57] calls the *late arrival with - delayed access*, i.e. arrivals occur just before the end of an epoch, services terminate at the end of an epoch and the arriving customer is blocked from entering an empty service facility until the servicing interval terminates. This is discrete time analysis of queuing systems as proposed by Neuts and Dafermos [35], Neuts and Klimko [83], Klimko and Neuts [64], Galliher and Wheeler [44] and Minh [78]. We are simply taking a snapshot of the system at these time epochs. We will not consider the other two discrete time arrangements described by Hunter [57] and Takagi [94], since the modeling aspects will be repetitious with only minor modifications. Hunter [57] explained three possible arrangements of arrival and service processes in discrete time. They are: 1) Early arrival 2) Late arrival with immediate access and 3) Late arrival with delayed access. In this book we work with Hunter's Case 1) i.e. Late arrival wth delayed access. The other two types of arrangements of discrete time views described by Hunter will not be considered in this book. However, it is very simple to extend the discussions in this book to those other two arrangements.

Chapter 2
Markov Processes

2.1 Introduction

Markov processes are a special class of stochastic processes. In order to fully understand Markov processes we first need to introduce stochastic processes.

2.2 Stochastic Processes

Consider a sample space $\mathscr{S}$ and a random variable $X(\omega)$ where $\omega \in \mathscr{S}$. A stochastic process may be defined as a collection of points $X(t,\omega),\ t \geq 0$, indexed over the parameter t, usually representing time but may represent others such as space, etc. Put in a simpler form we can say a stochastic process is a collection of random variables $X(t),\ t \geq 0$, which is observed at time t. Examples of a stochastic process could be the amount of rainfall $X(t)$ on day t, the remaining gasoline in the tank of an automobile at any time, etc. Most quantities studied in a queueing system involve a stochastic process. We will introduce the concept in the context of discrete time queues in the next section.

In this section, we consider integer valued stochastic processes in discrete times. Throughout this book we only consider non-negative integer valued processes. Consider a state space $\mathscr{I} = \{0,1,2,3,\cdots\}$ and a state space $\mathscr{I}_s$ which is a subset of $\mathscr{I}$, i.e. $\mathscr{I}_s \subseteq \mathscr{I}$. Now consider a random variable X_n which is observable at time $n = 0,1,2,\cdots$ in discrete time, where $X_n \in \mathscr{I}_s$.

The collection of these $X_0, X_1, X_2, \cdots$ form a stochastic process in discrete time; or written as $\{X_n, n = 0,1,2,\cdots\}$. A discrete parameter stochastic process is simply a family of random variables $X_n,\ n = 0,1,2,\cdots$ which evolves with parameter n, and in this case n could be time or any other quantity, such as an event count. This collection describes how the system evolves with time or event. Our interest is to study the relations of these collections. For example X_n could be the number of data packets that arrive at a router at time $n,\ n = 0,1,2,\cdots$, or the number of packets left

DOI 10.1007/978-1-4419-7314-6_2,

behind in the router by the nth departing packet. Examples of stochastic processes usually of interest to us in queueing are the number of packets in a queue at a particular time or between short intervals of time, the waiting time of packets which arrive in the system at particular times or between intervals of time, the waiting time of the nth packet, etc.

Of interest to us about a stochastic process is usually its probabilistic aspects; the probability that X_n assumes a particular value. Also of interest to us is the behaviour of a collection of a group of the random variables. For example the $Pr\{X_m = a_m, X_{m+1} = a_{m+1}, \cdots, X_{m+v} = a_{m+v}\}$. Finally we may be interested in the moments of the process at different times.

As an example, consider a communication processing system to which messages arrive at time $A_1, A_2, A_3, \ldots$, where the nth message arrives at time A_n. Let the nth message require S_n units of time for processing. We assume that the processor can only serve one message at a time in a first come first served order. Let there be an infinite waiting space (buffer) for messages waiting for service. Let X_t be the number of messages in the system at time t. X_t is a stochastic process. Further let D_n be the departure time of the nth message from the system after service completion.

In order to understand all the different ways in which this stochastic process can be observed in discrete time, we consider a particular realization of this system. For example, let us consider the process for up to 10 message arrivals for a particular realization as shown in Table (2.1) We can now study this process at any point in

n	A_n	S_n	D_n
1	0	3	3
2	2	4	7
3	5	1	8
4	6	3	11
5	7	6	17
6	12	2	19
7	14	4	23
8	15	5	28
9	19	1	29
10	25	3	32
⋮	⋮	⋮	⋮

Table 2.1 Example of a sample path

time.

In discrete time, we observe the system at times $t_1, t_2, t_3, \ldots$, where for simplicity, without loss of generality, we set $t_1 = 0$ to be the time of the first arrival, we have several choices of the values assumed by t_i, $i > 1$ in that we can study the system at:

- Equally spaced time epochs such that $t_2 - t_1 = t_3 - t_2 = t_4 - t_3 = \ldots = \tau$. For example, if $\tau = 1$, then $(t_1, t_2, t_3, \ldots) = (0, 1, 2, \ldots)$. In which case, $X_{t_1} = X_0 = 1$, $X_{t_2} = X_1 = 1, X_{t_3} = X_2 = 2, X_{t_4} = X_3 = 1, X_{t_5} = X_4 = 1, X_{t_6} = X_5 = 2, X_{t_7} = X_6 = 3, \cdots$

- Arbitrarily spaced time epochs such that $t_2 - t_1 = \tau_1$, $t_3 - t_2 = \tau_2$, $t_4 - t_3 = \tau_3$, . . . with τ_1, τ_2, τ_3, . . . not necessarily being equal. For example, if $\tau_1 = 1$, $\tau_2 = 2$, $\tau_3 = 1$, $\tau_4 = 4$, . . . hence $(t_1,t_2,t_3,t_4,t_5,\cdots) = (0,1,3,4,8,\cdots)$ we have $X_{t_1} = X_0 = 1, X_{t_2} = X_1 = 1, X_{t_3} = X_3 = 1, X_{t_4} = X_4 = 1, X_{t_5} = X_8 = 2,\cdots$
- Points of event occurrences. For example if we observe the system at departure times of messages, i.e. $(t_1,t_2,t_3,...) = (D_1,D_2,D_3,...) = (3,7,8,\cdots)$. In this case we have $X_{t_1} = X_3 = 1, X_{t_2} = X_7 = 3, X_{t_3} = X_8 = 2,\cdots$

Any events that occur between the time epochs of our observations are not of interest to us in discrete time analysis. We are only taking "snap shots" of the system at the specific time epochs.

By this assumption, then any discrete time observation is contained in continuous time observations.

Note that we may also have multivariate stochastic processes. For example, let X_n be the number of packets in the system at time n and let $Y_n \le X_n$ be how many of the X_n packets require special service. The stochastic process for this system is $\{X_n, Y_n, n = 01,2,3\cdots\}$, $X_n \in \mathscr{I}$, $Y_n \le X_n$. We will discuss multivariate processes later when we consider Markov chains.

2.3 Markov Processes

In 1907, A. A. Markov published a paper [75] in which he defined and investigated the properties of what are now called as Markov processes. A Markov process is a special stochastic process in which the state of the system in the future is independent of the past history of the system but dependent on only the present. The consideration of Markov processes is very important and in fact, is central to the study of queuing systems because one usually attempts to reduce the study of nearly all queueing systems to the study of Markov process.

Consider a stochastic process X_n in discrete time $(n = 0,1,2,...)$. The set of possible values that X_n takes on is referred as its state space. X_n is a Markov process if

$$\begin{aligned} &Pr\{X_{n+1} \in A_{n+1} | X_n = x_n, X_{n-1} \in A_{n-1}, ..., X_0 \in A_0\} \\ &= Pr\{X_{n+1} \in A_{n+1} | X_n = x_n\} \end{aligned}$$

holds for each state integral n, each state x_n and the set of states A_{n+1}, A_{n-1}, A_0.

If the numbers in the state space of a Markov process are finite or countable, the Markov process is called a Markov chain. In this case, the state space is usually assumed to be the set of integers $\{0, 1, \ldots\}$. In the case that the numbers in the state space of a Markov process are over a finite or infinite continuous interval (or set of such intervals), the Markov process is referred to as a Markov process. For most times in queueing applications, Markov chains are more common even though continuous space Markov processes, in the form of stochastic fluid models, are getting popular for applications to queueing systems. We focus mainly on Markov chains in this chapter.

2.4 Discrete Time Markov Chains

Consider a discrete time stochastic process X_0, X_1, . . ., with discrete (i.e. finite or countable) state space $\mathscr{I}_c = i_0, i_1, i_2, ... \subseteq \mathscr{I}$. If

$$\begin{aligned} &Pr\{X_{n+1} = i_{n+1} | X_n = i_n, X_{n-1} = i_{n-1}, ..., X_0 = i_0\} \\ &= Pr\{X_{n+1} = i_{n+1} | X_n = i_n\} \end{aligned}$$

holds for any integral n, and the states i_{n+1}, i_n, i_{n-1}, . . ., i_0, then X_n is said to be a discrete time Markov chain (DTMC). If further we have

$$Pr\{X_{n+m+1} = j | X_{n+m} = i\} = Pr\{X_{n+1} = j | X_n = i\}, \forall (i, j) \in \mathscr{I}, \forall (m, n) \geq 0,$$

we say the Markov chain is time-homogeneous or stationary. Throughout most of this book we shall be dealing with this class of DTMC.

Generally we say

$$p_{i,j}(n) = P\{X_{n+1} = j | X_n = i\}$$

is the transition probability from state i at time n to state j at time $n+1$. However, for most of our discussions in this book (until later chapters) we focus on the case where the transition is independent of the time, i.e. $p_{i,j}(n) = p_{i,j}$, $\forall n$. The matrix $P = (p_{i,j}), (i, j) \in \mathscr{I}$ is called the transition probability matrix of the Markov chain, in which

$$p_{i,j} = Pr\{X_{n+1} = j | X_n = i\}.$$

The basic properties of the elements of the transition probability matrix of a Markov chain are as follows:

Properties:

$$0 \leq p_{i,j} \leq 1, \quad \forall i \in \mathscr{I}, \forall j \in \mathscr{I}$$
$$\sum_{j \in \mathscr{I}} p_{ij} = 1$$

For example consider a DTMC with the state space $\{1, 2\}$. We show, in Fig (2.1) below, its transition behaviour between time n and $n+1$.

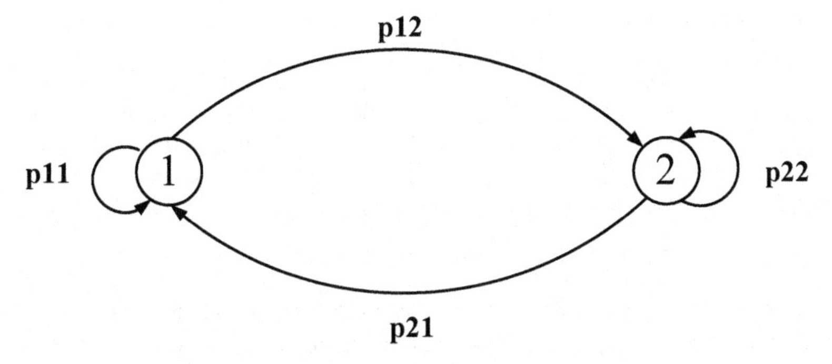

Fig. 2.1 Two States Transition Probabilities

2.4.1 Examples

1. Example 2.1.1: Consider a discrete stochastic process X_n which is the number of data packets in the buffer of a router at time n. At time n one packet is removed from the buffer if it is not empty and A_n packets are added at the same time. Let $\{A_n,\ n = 0, 1, 2, \cdots\}$ be a discrete stochastic process which assumes values in the set $\mathscr{I}$. This DTMC can be written as

$$X_{n+1} = (X_n - 1)^+ + A_n,\ \ n = 0, 1, 2, 3, \cdots,$$

where $(z)^+ = max\{z, 0\}$. For example, let $A_n = A,\ \forall n$ and $a_k = Pr\{A = k\},\ k = 0, 1, 2, 3, \cdots$, then

$$p_{0,j} = a_j,\ j = 0, 1, 2, \cdots,$$

$$p_{i,j} = a_{j-i+1}, 1 \leq i \leq j = 1, 2, 3, \cdots,$$

and

$$p_{i,j} = 0,\ j < i.$$

The transition matrix P associated with this DTMC can be written as

$$P = \begin{bmatrix} a_0 & a_1 & a_2 & a_3 & a_4 & \cdots \\ a_0 & a_1 & a_2 & a_3 & a_4 & \cdots \\ & a_0 & a_1 & a_2 & a_3 & \cdots \\ & & a_0 & a_1 & a_2 & \cdots \\ & & & \ddots & \ddots & \ddots \end{bmatrix}.$$

2. Example 2.1.2: Use the same example as above but put a limit on how many packets can be in the buffer. If we do not allow the number to exceed $K < \infty$, then this DTMC can be written as

$$X_{n+1} = min\{(X_n - 1)^+ + A_n, K\}\ \ n = 0, 1, 2, 3, \cdots.$$

For this system we have

$$p_{0,j} = a_j, \; j = 0,1,2,\cdots,K-1, \;\; p_{0,K} = \tilde{a}_K,$$

$$p_{i,j} = a_{j-i+1}, 1 \le i \le j = 1,2,3,\cdots,K-1, \;\; p_{i,K} = \tilde{a}_{K-i+1},$$

and

$$p_{i,j} = 0, \; j < i,$$

where $\tilde{a}_w = \sum_{v=w}^{\infty} a_v$.

3. Example 2.1.3: Consider a discrete stochastic process Y_n which is the number of packets in a buffer at time n. Suppose at time n, one packet is added to contents of the buffer and the minimum of $Y_n + 1$ and B_n items are removed from the buffer, where $\{B_n, \; n = 0,1,2,\cdots\}$ is a discrete stochastic process which assumes values in the set $\mathscr{I}$. This DTMC can be written as

$$Y_{n+1} = (Y_n + 1 - B_n)^+, \;\; n = 0,1,2,3,\cdots,$$

where $(z)^+ = max\{z,0\}$. For example, let $B_n = B, \; \forall n, \; b_k = Pr\{B = k\}, \; k = 0,1,2,3,\cdots$, and $\tilde{b}_k = \sum_{v=k+1}^{\infty} b_v$, then

$$p_{i,0} = \tilde{b}_i, \; i = 0,1,2,\cdots,$$

$$p_{i,j} = b_{i+1-j}, 1 \le j \le i = 1,2,3,\cdots,$$

and

$$p_{i,j} = 0, \; i < j.$$

The transition matrix P associated with this DTMC can be written as

$$P = \begin{bmatrix} \tilde{b}_0 & b_0 & \cdots & & \\ \tilde{b}_1 & b_1 & b_0 & \cdots & \\ \tilde{b}_2 & b_2 & b_1 & b_0 & \cdots \\ \vdots & \vdots & \vdots & \vdots & \ddots \end{bmatrix}.$$

4. Example 2.1.4: Consider two types of items; type A and type B. There are N of each, i.e. a total of $2N$ items. Suppose these $2N$ items are randomly divided into two sets of equal numbers with N placed in one urn and the remaining N placed in the other urn. If there are X_n of type A items in the first urn at time n then we have $N - X_n$ of type B in the first urn. If we now take one item each at random from each urn and exchange their urns, then it is clear that $X_n, \; n = 0,1,2,3,\cdots$ is a represented as

$$X_{n+1} = \begin{cases} X_n - 1, & \text{with probability } \frac{X_n^2}{N^2} \\ X_n, & \text{with probability } \frac{2X_n(N-X_n)}{N^2} \\ X_n + 1, & \text{with probability } \frac{(N-X_n)^2}{N^2} \end{cases}$$

This is the well known Bernoulli-Laplace diffusion model. We can write

$$p_{i,i-1} = (\frac{i}{N})^2,$$

$$p_{i,i} = 2\frac{i}{N}(1-\frac{i}{N}),$$

$$p_{i,i+1} = (1-\frac{i}{N})^2.$$

For the example of when $N = 4$ we can write the transition matrix as

$$P = \begin{bmatrix} 0 & 1 & & & \\ \frac{1}{16} & \frac{3}{8} & \frac{9}{16} & & \\ & \frac{1}{4} & \frac{1}{2} & \frac{1}{4} & \\ & & \frac{9}{16} & \frac{3}{8} & \frac{1}{16} \\ & & & 1 & 0 \end{bmatrix}.$$

5. Example 2.1.5: Consider a single server queueing system in which arrivals and service of packets are independent. Let A_n be the interarrival time between the nth and the $n-1^{st}$ packet, S_n the processing time of the nth packet and W_n the waiting time of the nth packet, then we can see that W_n, $n = 1,2,3,\cdots$ is a Markov chain given as
$$W_{n+1} = (W_n + S_n - A_{n+1})^+.$$
Here the parameter n is not time but a sequential label of a packet.
6. Example 2.1.6: The Gilbert-Elliot (G-E) channel model is a simple, but effective two-state DTMC. Consider a communication channel which is subject to noise. The G-E model assumes that at all times the channel is in two possible states: state 0 – channel is good, and state 1 – channel is bad. The state of the channel at the next time slot depends only on the state at the current time slot and not on the previous times. The probability of the channel state going from good to bad is q and probability of it going from bad to good is p. With this system as a DTMC with state space $\{0,1\}$, its transition matrix is given as
$$P = \begin{bmatrix} 1-q & q \\ p & 1-p \end{bmatrix}.$$
7. Example 2.1.7: The Fritchman channel model is an extension of the G-E model in that it assumes that the channel could be in more than 2 possible states. Let the number of states be M with $2 < M < \infty$, which are numbered $\{0,1,2,\cdots,M-1\}$. We let state 0 correspond to the best channel state and state $M-1$ to the worst channel state. The Fritchman model restricts the channel changes from one time to the next time not to jump up or down by more than one state. The probability of going from state i to $i+1$ is given as u and going from i to $i-1$ is given as d, while that of remaining in i is given as a. It also assumes that the channel state at the next time depends only on the current state. The transition matrix for this

model is written as

$$P = \begin{bmatrix} 1-u & u & & & & \\ d & a & u & & & \\ & d & a & u & & \\ & & \ddots & \ddots & \ddots & \\ & & & d & a & u \\ & & & & d & 1-d \end{bmatrix}.$$

2.4.2 State of DTMC at arbitrary times

One of our interests in DTMC is the state of the chain at an arbitrary time, i.e. $Pr\{X_n = i\}$. This we can obtain by using the total probability rule as

$$Pr\{X_{n+1} = i\} = \sum_{k \in \mathscr{I}} Pr\{X_{n+1} = i | X_n = k\} Pr\{X_n = k\}. \tag{2.1}$$

Let $\boldsymbol{x}^{(n)} = [x_0^{(n)}, x_1^{(n)}, ...,]$ be the state probability vector describing the state of the system at time n, where $x_i^{(n)}$ = Probability that the system is in the state i at time n. Then we can write

$$x_i^{(n+1)} = \sum_{k \in \mathscr{I}} x_k^{(n)} p_{k,i} \tag{2.2}$$

and in matrix form we have

$$\boldsymbol{x}^{(n+1)} = \boldsymbol{x}^{(n)} P = \boldsymbol{x}^{(0)} P^n. \tag{2.3}$$

Consider a DTMC with the state space $\{1,2\}$, then the transition probabilities and the state of the system at any arbitrary times are shown in the Fig (2.1).

Another way to view this process is to draw N nodes directed network as shown in Fig (2.2) below, for the case of $N = 2$, and then connect the node pairs (i, j) that have $p_{ij} > 0$. If we now consider the case in which the DTMC goes from node i to node j with the probability p_{ij} at each time step, then $x_i^{(n)}$ is the probability that we find the DTMC at node i at time n.

We now introduce some examples as follows:

2.4.2.1 Example

Example 2.2.1: Consider a processor which serves messages one at a time in order of arrival. Let there be a finite waiting buffer of size 4 for messages. Assume messages arrive according to the Bernoulli process with parameter a ($\bar{a} = 1 - a$) and service is geometric with parameter b ($\bar{b} = 1 - b$). A Bernoulli process, to be discussed later, is a stochastic process with only two possible outcomes, viz: success

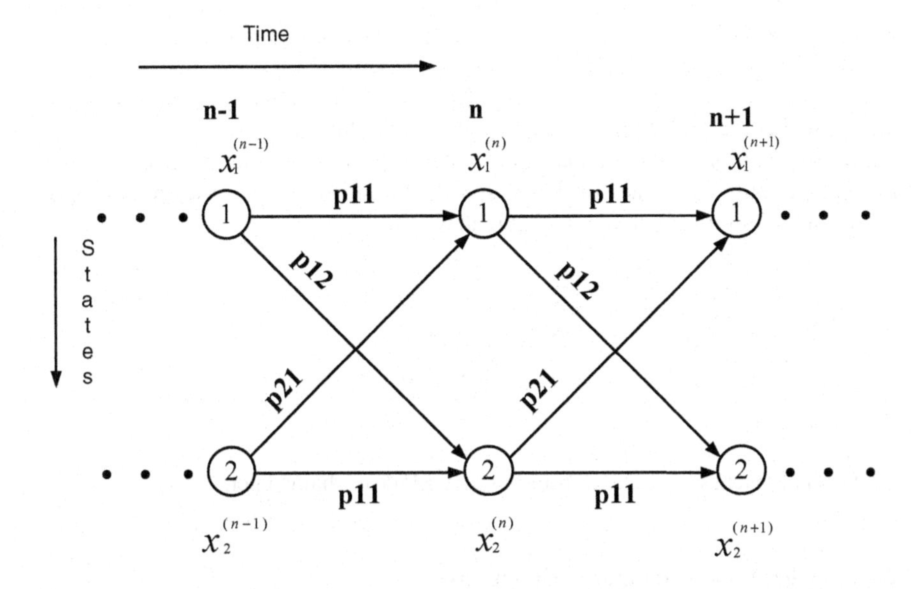

Fig. 2.2 State Probabilities at Arbitrary Times

or failure, with each trial being independent of the previous ones. The parameter a represents the probability of a success and $\bar{a} = 1 - a$ represents the probability of a failure. A geometric distribution describes the number of trials between successive successes. These will be discussed in more details in Chapter 3. Let X_n be the number of messages in the system, including the one being served, at time n. X_n is a DTMC, with transition matrix P given as

$$P = \begin{bmatrix} \bar{a} & a & & & \\ \bar{a}b & \bar{a}\bar{b}+ab & a\bar{b} & & \\ & \bar{a}b & \bar{a}\bar{b}+ab & a\bar{b} & \\ & & \bar{a}b & \bar{a}\bar{b}+ab & a\bar{b} \\ & & & \bar{a}b & \bar{a}\bar{b}+a \end{bmatrix}$$

If $a = 0.3$ and $b = 0.6$, then we have

$$P = \begin{bmatrix} 0.70 & 0.30 & & & \\ 0.42 & 0.46 & 0.12 & & \\ & 0.42 & 0.46 & 0.12 & \\ & & 0.42 & 0.46 & 0.12 \\ & & & 0.42 & 0.58 \end{bmatrix}$$

If at time zero the system was empty, i.e. $\boldsymbol{x}^{(0)} = [1,0,0,0,0]$, then $\boldsymbol{x}^{(1)} = \boldsymbol{x}^{(0)}P = [0.7,0.3,0.0,0.0,0.0]$. Similarly, $\boldsymbol{x}^{(2)} = x^{(1)}P = x^{(0)}P^2 = [0.616,0.348,0.036,0.0,0.0]$

and $\boldsymbol{x}^{(5)} = \boldsymbol{x}^{(4)}P = \boldsymbol{x}^{(0)}P^5 = [0.5413, 0.3639, 0.0804, 0.0130, 0.0014]$. We also obtain $\boldsymbol{x}^{(100)} = \boldsymbol{x}^{(99)}P = \boldsymbol{x}^{(0)}P^{100} = [0.5017, 0.3583, 0.1024, 0.0293, 0.0084]$. We notice that $x^{(101)} = \boldsymbol{x}^{(100)}P = \boldsymbol{x}^{(0)}P^{101} = [0.5017, 0.3583, 0.1024, 0.0293, 0.0084] = \boldsymbol{x}^{(100)}$, in this case. We will discuss this observation later.

Keeping in mind that a Markov chain can also be of infinite state space, consider this same example but where the buffer space is unlimited. In that case, the state size is countably infinite and the transition matrix P is given as

$$P = \begin{bmatrix} \bar{a} & a & & & \\ \bar{a}b & \bar{a}\bar{b} + ab & a\bar{b} & & \\ & \bar{a}b & \bar{a}\bar{b} + ab & a\bar{b} & \\ & & \bar{a}b & \bar{a}\bar{b} + ab & a\bar{b} \\ & & & \ddots & \ddots \quad \ddots \end{bmatrix}$$

We shall discuss the case of infinite space Markov chains later.

2.4.2.2 Chapman Kolmogorov Equations

Let $P^{(n)}$ be the matrix whose (i,j)th entry refers to the probability that the system is in state j at the nth transition, given that it was in state i at the start (i.e. at time zero), then $P^{(n)} = P^n$, i.e., the matrix $P^{(n)}$ is the nth order transition matrix of the system. In fact, let $p_{ij}^{(n)}$ be the (i,j)th element of $P^{(n)}$ then we have

$$\begin{aligned} p_{ij}^{(2)} &= \sum_{v \in \mathscr{I}} p_{iv} p_{vj} \\ p_{ij}^{(3)} &= \sum_{v \in \mathscr{I}} p_{iv}^{(2)} p_{vj} = \sum_{v \in \mathscr{I}} p_{iv} p_{vj}^{(2)} \\ \vdots &= \vdots \\ p_{ij}^{(n)} &= \sum_{v \in \mathscr{I}} p_{iv}^{(n-1)} p_{vj} = \sum_{v \in \mathscr{I}} p_{iv} p_{vj}^{(n-1)} \end{aligned}$$

which when written out in a more general form becomes

$$p_{ij}^{(n)} = \sum_{v \in \mathscr{I}} p_{iv}^{(m)} p_{vj}^{(n-m)}, \tag{2.4}$$

and in matrix form becomes

$$P^{(n)} = P^{(n-m)} P^{(m)}, \text{ i.e., } P_{ij}^{(n)} = \sum_{v \in \mathscr{I}} P_{iv}^{(m)} p_{vj}^{(n-m)} \tag{2.5}$$

Since $P^{(2)} = P^2$ from the first of above equation, by a simple induction, we will know $P^{(n)} = P^n$ holds for any non-negative integer n. This is the well known Chapman-Kolmogorov equation.

2.4.3 Classification of States

In this section, we explain how the states of a Markov chain are classified. Consider a 9 state DTMC shown in Fig(2.3) below. We let an arrow from i to j imply that $p_{ij} > 0$, and assume that $p_{ii} > 0$ for all the 9 states. We use this figure to help us in explaining the classes of states of a DTMC.

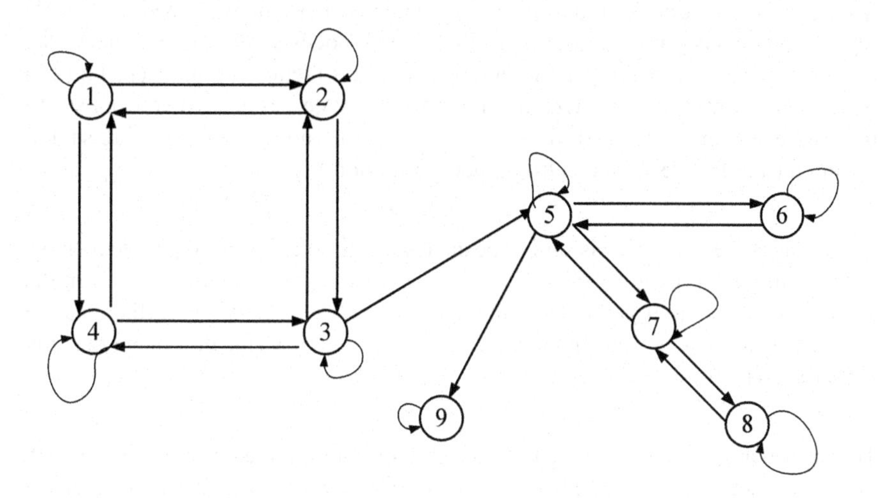

Fig. 2.3 Markov Chain Network Example

Communicating States - State j is said to be accessible from state i if there exists a finite n such that $p^n_{i,j} > 0$. If two states i and j are accessible to each other, they are said to communicate. Consider the nine state DTMC with state space $\{1,2,3,4,5,6,7,8,9\}$ and the following transition matrix

$$P = \begin{bmatrix} p_{1,1} & p_{1,2} & & p_{1,4} & & & & & \\ p_{2,1} & p_{2,2} & p_{2,3} & & & & & & \\ & p_{3,2} & p_{3,3} & p_{3,4} & p_{3,5} & & & & \\ p_{4,1} & & p_{4,3} & p_{4,4} & & & & & \\ & & & & p_{5,5} & p_{5,6} & p_{5,7} & & p_{5,9} \\ & & & & p_{6,5} & p_{6,6} & & & \\ & & & & p_{7,5} & & p_{7,7} & p_{7,8} & \\ & & & & & & p_{8,7} & p_{8,8} & \\ & & & & & & & & p_{9,9} \end{bmatrix}.$$

We assume that all the cells with entries have values that are greater than zero. For example, $p_{1,2} > 0$, whereas, $p_{1,3} = 0$. For example in this DTMC states 1 and 2 do communicate, whereas state 4 is not accessible from state 5 but state 5 is accessible from state 3. This transition matrix can also be shown as a network with each node representing a state and a directed arc joining two nodes signify a non-zero value, i.e. direct access (transition) between the two nodes (states) in one step. Fig(2.3) represents this transition matrix. In general we say state j is accessible from state i if a path exists in the network from node i to node j, and the two states are said to communicate if there are paths from node $i(j)$ to node $j(i)$.

Absorbing States - a state i is said to be an absorbing state if once the system enters state i it never leaves it. Specifically, state i is an absorbing state if $p_{ii} = 1$. In the example above state 9 is an absorbing state. Also if we set $p_{6,5} = 0$, then state 6 becomes an absorbing state and as we can see $p_{6,6} = 1$ by virtue of it being the only non-zero element in that row.

Transient States - a state i is said to be a transient state if there exists a state j which can be reached in a finite number of steps from state i but state i cannot be reached in a finite number of steps from state j. Specifically, state i is transient if there exists a state j such that $p_{ij}^{(n)} > 0$ for some $n < \infty$ and $p_{ji}^{(m)} = 0, \ \forall m > 0$. In the above example, state 5 can be reached from state 3 in a finite number of transitions but state 3 can not be reached from state 5 in a finite number of transitions, hence state 3 is a transient state.

Recurrent States - a state i is said to be recurrent if the probability that the system ever returns to state i is 1. Every state which is not a transient state is a recurrent state. Consider a three state DTMC with states $\{0, 1, 2\}$ and probability transition matrix P given as

$$P = \begin{bmatrix} 0.2 & 0.7 & 0.1 \\ 0.1 & 0.3 & 0.6 \\ & 0.6 & 0.4 \end{bmatrix}.$$

All the states here are recurrent. If however, the matrix P is

$$P = \begin{bmatrix} 0.2 & 0.8 & \\ 0.1 & 0.3 & 0.6 \\ & & 1 \end{bmatrix},$$

then states 0 and 1 are transient with state 2 being an absorbing state and technicaly also being a recurrent state. Later we show that there are two types of recurrent states, viz: positive recurrent for which the mean recurrence time is finite and the null recurrent for which the mean recurrence time is infinite.

Periodic States - a state i is said to be periodic if the number of steps requireD to return to it is a multiple of k, $(k > 1)$. For example, consider a Markov chain with the following transition matrix

$$P = \begin{bmatrix} 0 & 1 & 0 \\ 0 & 0 & 1 \\ 1 & 0 & 0 \end{bmatrix},$$

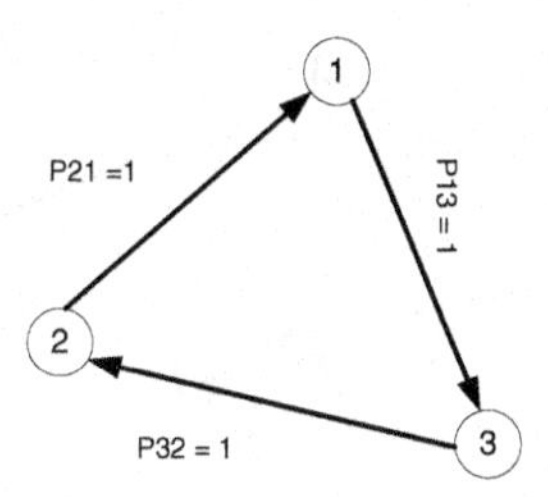

Fig. 2.4 Periodic States Example

This chain is also represented as a network in Fig(2.4). All the three states are periodic with period of $k = 3$.

If a state is not periodic, it is said to be aperiodic. In Fig(2.4), to go from any state back to the same state requires hopping over three links, i.e. visiting two other nodes all the time.

2.4.4 Classification of Markov Chains

In this section, we show how Markov chains are classified.

Irreducible Chains - a Markov chain is said to be irreducible if all the states in the chain communicate with each other. As an example consider two Markov chains with five states S_1, S_2, S_3, S_4, S_5 and the following transition matrices

$$P_1 = \begin{bmatrix} 0.70 & 0.30 & & & \\ 0.40 & 0.30 & 0.15 & 0.15 & \\ & 0.60 & 0.10 & 0.30 & \\ & & 0.30 & 0.50 & 0.20 \\ & & & 0.20 & 0.80 \end{bmatrix} \qquad P_2 = \begin{bmatrix} 0.70 & 0.30 & & & \\ 0.40 & 0.60 & & & \\ & & 0.10 & 0.90 & \\ & & 0.30 & 0.50 & 0.20 \\ & & & 0.20 & 0.80 \end{bmatrix}.$$

The DTMC with transition matrix P_1 is irreducble, but the one with P_2 is reducible into two sub-sets of states: the set of states in $A = \{S1, S2\}$ and the set of states in $B = \{S3, S4, S5\}$. The set of states in A do not communicate with the set of states in B. Hence, this is a reducible Markov chain. The set of states in A are said to form a closed set and so also are the states in B.

Absorbing Chains - a Markov chain is said to be an absorbing chain if any of its states is an absorbing state, i.e. if at least for one of its states i we have $Pr\{X_{n+1} = i|X_n = i\} = p_{i,i} = 1$. As an example consider an absorbing DTMC with states $\{0, 1, 2\}$ in which state 0 is an absorbing state. Fig(2.5) is a an example representation of such a chain.

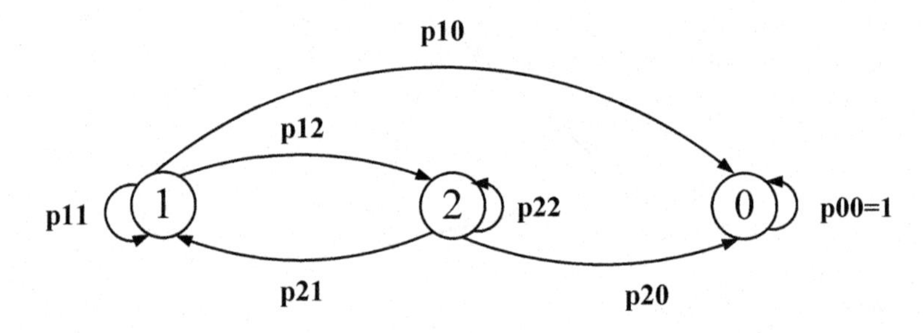

State 0 is the absorbing state

Fig. 2.5 An Absorbing Markov Chain

Recurrent Markov Chain - a Markov chain is said to be recurrent if all the states of the chain are recurrent. It is often important to also specify if a chain is positive or null recurrent type based on if any of the states are positive recurrent or null recurrent. The conditions that govern the differences between these two recurrence will be presented later in the next subsection. Proving that a Markov chain is positive recurrent is usually involved and problem specific. We will not attempt to give a general method for proving recurrence in this book. However, we will present the method used for proving recurrence for some classes of Markov chains which are commonly encountered in queueing theory.

Ergodic Markov Chain - an irreducible Markov chain is said to be ergodic if all its states are aperiodic and positive recurrent. In addition, if the chain is finite, then aperiodicity is all it needs to be an ergodic chain.

2.5 First Passage Time

In this section we consider first passage time, a very important measure which is used in many situations to characterize a Markov chain. Consider a Markov chain with the state space $\mathscr{N} = 0,1,2,...,N$, $\mathscr{N} \subseteq \mathscr{I}$ and associated transition matrix P. Given that the chain is in a particular state, say i at any arbitrary time, we want to find from then on when it reaches a state j for the first time. Define $f_{ij}^{(n)}$ as the probability that state j is visited for the first time at the nth transition, given that the system was in state i at the start, i.e. at time 0. When $j = i$ we have $f_{ii}^{(n)}$ as the probability that state i is visited for the first time at the nth transition, given that the system was in state i at the start, i.e. at time 0; this is known as first return probability. Note that $p_{ij}^{(n)}$ is different from $f_{ij}^{(n)}$ in that the former gives the probability that the system is in state j at the nth transition given that it was in state i at the start; the system could have entered state j by the 1st, 2nd, . . ., or $(n-1)$th transition and remained in, or went into another state and returned to j by the nth transition. Note that $p_{ij}^{(n)} \geq f_{ij}^{(n)}$, $\forall n$. Consider a DTMC with states $\{1,2,3\}$ we show in Figs (2.6) and (2.7) below, the difference between $f_{2,3}^{(2)}$ and $p_{2,3}^{(2)}$.

We see that

$$p_{2,3}^{(2)} = f_{2,3}^{(2)} + p_{2,3}p_{3,3}.$$

It is straightforward to see that

$$p_{ij}^{(n)} = f_{ij}^{(n)} + (1-\delta_{n1})\sum_{v=1}^{n-1} f_{ij}^{(v)} p_{jj}^{(n-v)}, \ n \geq 1,$$

where δ_{ij} is the kronecker delta, i.e. $\delta_{ij} = \begin{cases} 1, & i=j \\ 0, & \text{otherwise} \end{cases}$.

The arguments leading to this expression are as follows: The system is in state j at the nth transition, given that it started from state i at time zero in which case, either

1. the first time that state j is visited is at the nth transition - this leads to the first term, or

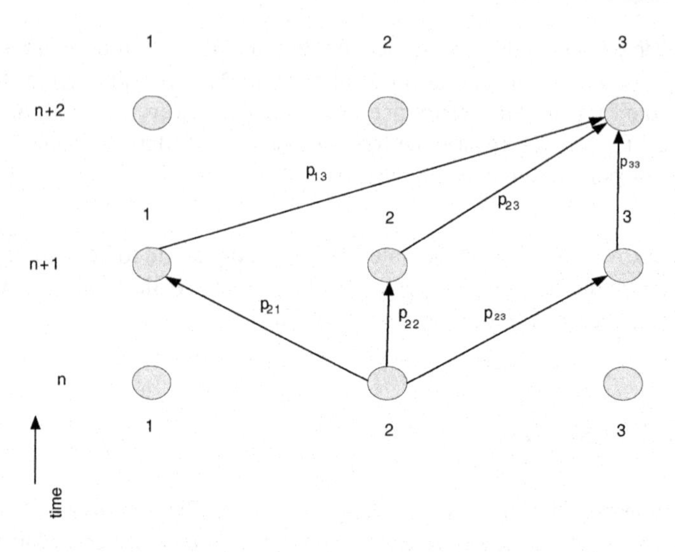

Fig. 2.6 Case of Two Step Probability

2. the first time that state j is visited is at the vth transition ($v < n$) and then for the remaining $n - v$ transitions the system could visit the j state as many times as possible (without visiitng state i in between) but must end up in state j by the last transition.

After rewriting this equation, we obtain

$$f_{ij}^{(n)} = p_{ij}^{(n)} - (1 - \delta_{n1}) \sum_{v=1}^{n-1} f_{ij}^{(v)} p_{jj}^{(n-v)}, \ n \geq 1 \tag{2.6}$$

Equation (2.6) is applied recursively to calculate first passage time probability. For example, to determine $f_{ij}^{(T)}$, $T > 1$, we initiate the recursion by computing $p_{ij}^{(n)} \geq f_{ij}^{(n)}$, $\forall n$. It is straightforward to note that

$$f_{ij}^{(1)} = p_{ij}^{(1)}$$

and then calculate

$$f_{ij}^{(2)} = p_{ij}^{(2)} - f_{ij}^{(1)} p_{jj}^{(1)}$$
$$f_{ij}^{(3)} = p_{ij}^{(3)} - \sum_{v=1}^{2} f_{ij}^{(v)} p_{jj}^{(3-v)}$$

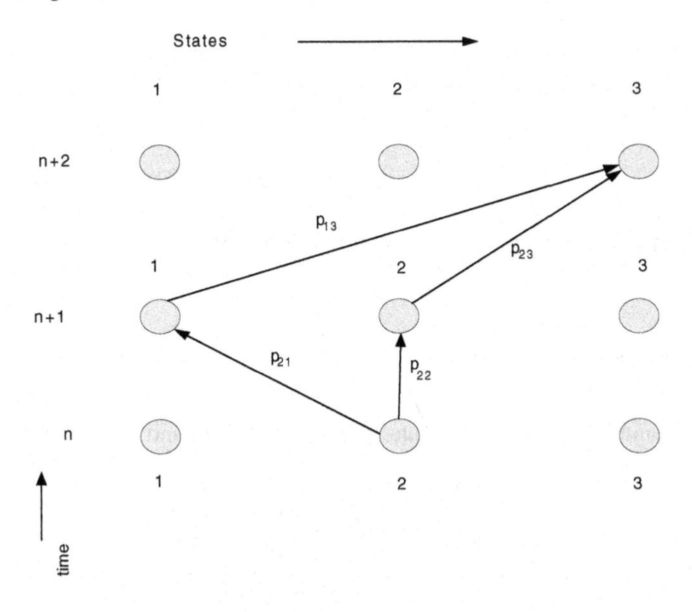

Fig. 2.7 Case of First Passage Probability

$$\vdots = \vdots$$
$$f_{ij}^{(T)} = p_{ij}^{(T)} - \sum_{v=1}^{T-1} f_{ij}^{(v)} p_{jj}^{(T-v)}$$

Note that we have to calculate P^n, $n = 2, 3, ..., T-1$, during the recursion by this method. However, Neuts [85] has derived a more efficient procedure for computing the first passage probability and is given as:

$$f_{ij}^{(n+1)} = \sum_{v \in \mathscr{I}} p_{iv} f_{vj}^{(n)} - p_{ij} f_{jj}^{(n)}, \; n \geq 1. \tag{2.7}$$

The arguments leading to this result are as follows:

$$\begin{aligned}
f_{ij}^{(2)} &= \sum_{v \in \mathscr{N}, v \neq j} p_{iv} f_{vj}^{(1)} = && \sum_{v \in \mathscr{N}} p_{iv} f_{vj}^{(1)} - p_{i,j} f_{jj}^{(1)} \\
f_{ij}^{(3)} &= \sum_{v \in \mathscr{N}, v \neq j} p_{iv} f_{vj}^{(2)} = && \sum_{v \in \mathscr{N}} p_{iv} f_{vj}^{(2)} - p_{i,j} f_{jj}^{(2)} \\
\vdots &= \qquad\qquad\quad \vdots = && \qquad\qquad\qquad\qquad \vdots \\
f_{ij}^{(n+1)} &= \sum_{v \in \mathscr{N}, v \neq j} p_{iv} f_{vj}^{(n)} = && \sum_{v \in \mathscr{N}} p_{iv} f_{vj}^{(n)} - p_{i,j} f_{jj}^{(n)}, \; n \geq 1.
\end{aligned}$$

If we define the matrix $F^{(n)}$ whose elements are $f_{ij}^{(n)}$, and $F_d^{(n)}$ as the matrix F with only its diagonal elements, i.e. $(F_d^{(n)})_{jj} = f_{jj}^{(n)}$ and $(F_d^{(n)})_{ij} = 0,\ i \neq j$, then the above equation can be written in matrix form as

$$F^{(n+1)} = PF^{(n)} - PF_d^{(n)} = P(F^{(n)} - F_d^{(n)}). \tag{2.8}$$

This is much easier to work with for computational purposes.

2.5.1 Examples

Consider Example 2.2.1 with the state space $\{0,1,2,3,4\}$ and transition matrix

$$P = \begin{bmatrix} 0.70 & 0.30 & & & \\ 0.42 & 0.46 & 0.12 & & \\ & 0.42 & 0.46 & 0.12 & \\ & & 0.42 & 0.46 & 0.12 \\ & & & 0.42 & 0.58 \end{bmatrix}.$$

Suppose we currently have 2 messages in the buffer, what is the probability that the buffer becomes full (i.e. buffer will reach state 4) for the first time at the fourth transition? We wish to determine $f_{2,4}^{(4)}$, i.e. the probability that the buffer becomes full for the first time at time 4, given that there were 2 in the system at time 0. Also, given that the buffer is now full, what is the probability that the next time that the buffer is full again is at the fifth transition from now? I let the reader figure this out as an excercise.

What we need to calculate is $F^{(4)}$ and select its $(2,4)$ element. By using the formula above we can actually obtain the whole matrix $F^{(4)}$ recursively as follows:

$$F^{(1)} = P = \begin{bmatrix} 0.70 & 0.30 & & & \\ 0.42 & 0.46 & 0.12 & & \\ & 0.42 & 0.46 & 0.12 & \\ & & 0.42 & 0.46 & 0.12 \\ & & & 0.42 & 0.58 \end{bmatrix}, \quad F_d^{(1)} = \begin{bmatrix} 0.70 & & & & \\ & 0.46 & & & \\ & & 0.46 & & \\ & & & 0.46 & \\ & & & & 0.58 \end{bmatrix}$$

We obtain

$$F^{(2)} = \begin{bmatrix} 0.1260 & 0.2100 & 0.0360 & 0.0 & 0.0 \\ 0.1932 & 0.1764 & 0.0552 & 0.0144 & 0.0 \\ 0.1764 & 0.1932 & 0.1008 & 0.0552 & 0.0144 \\ 0 & 0.1764 & 0.1932 & 0.1008 & 0.0552 \\ 0 & 0 & 0.1764 & 0.2436 & 0.0504 \end{bmatrix}.$$

After the fourth recursion we obtain

$$F^{(4)} = \begin{bmatrix} 0.0330 & 0.1029 & 0.0414 & 0.0070 & 0.0005 \\ 0.0701 & 0.0749 & 0.0362 & 0.0117 & 0.0024 \\ 0.1298 & 0.0701 & 0.0302 & 0.0200 & 0.0106 \\ 0.1022 & 0.1298 & 0.0726 & 0.0302 & 0.0200 \\ 0.0311 & 0.1111 & 0.1526 & 0.0819 & 0.0132 \end{bmatrix}.$$

Hence we have $f_{2,4}^{(4)} = 0.0106$.

2.5.2 Some Key Information Provided by First Passage

Define $f_{ij} = \sum_{n=1}^{\infty} f_{ij}^{(n)}$ as the probability of the system ever passing from state i to state j. It is clear that

- $f_{ij} = 1$ implies that state j is reachable from state i.
- $f_{ij} = f_{ji} = 1$ implies that states i and j do communicate
- $f_{ii}^{(n)}$ is the probability that the first recurrence time for state i is n. Letting $f_{ii} = \sum_{n=1}^{\infty} f_{ii}^{(n)}$ then
- $f_{ii} = 1$ for all recurrent states
- $f_{ii} < 1$ for all transient states.

Sometimes we are interested mainly in the first moment of the passage times. We start by obtaining the probability generating function of the first passage time. Let $P_{ij}(z) = \sum_{n=1}^{\infty} p_{ij}^{(n)} z^n$ and $F_{ij}(z) = \sum_{n=1}^{\infty} f_{ij}^{(n)} z^n$, $|z| < 1$ be the probability generating functions (*pgf*) of the probability of transition times and that of the first passage times, respectively. Taking the *pgf* of the first passage equation we obtain:

$$F_{ij}(z) = P_{ij}(z) - F_{ij}(z)P_{jj}(z) = \frac{P_{ij}(z)}{1+P_{jj}(z)}.$$

The nth factorial moment of the first passage time is then given as $\frac{d^n F_{ij}(z)}{dz^n}|_{z\to 1}$. However, this is not simple to evaluate unless $P_{ij}(z)$ has a nice and simple structure. The first moment can be obtained in an easier manner as follows.

The mean first passage time from state i to state j denoted by M_{ij} is given as

$$M_{ij} = \sum_{n=1}^{\infty} n f_{ij}^{(n)}.$$

Keeping in mind that when the system leaves state i it may go into state j directly with probability p_{ij} in which case it takes one unit of time, or it could go directly to another state $k(\neq j)$ in which case it takes one unit of time with probability p_{ik} plus the mean first passage time from state k to state j. Hence, we can calculate the mean first passage time as follows:

$$M_{ij} = p_{ij} + \sum_{k \neq j} (1 + M_{kj}) p_{ik} = 1 + \sum_{k \neq j} p_{ik} M_{jk}. \tag{2.9}$$

Define a matrix $M = (M_{ij})$. Let M_d be a diagonal matrix whose elements are $= M_{ii}$ with $\mathbf{1} = [1\ 1\ \cdots\ 1]^T$ and $\mathbf{e}_1 = [1\ 0\ 0\ \cdots\ 0]^T$, then the above equation can be written in matrix form as

$$M = \mathbf{1} \otimes \mathbf{1}^T + P(M - M_d),$$

where $\otimes$ is the kronecker product operator, i.e. for two matrices A of order $n \times m$ and B of order $u \times v$ we have $C = A \otimes B$ of dimension $nu \times mv$ with structure

$$C = \begin{bmatrix} A_{11}B & A_{12}B & \cdots & A_{1m}B \\ A_{21}B & A_{22}B & \cdots & A_{2m}B \\ \vdots & \vdots & \cdots & \vdots \\ A_{n1}B & A_{n2}B & \cdots & A_{nm}B \end{bmatrix},$$

where $A_{ij},\ i = 1,2,\cdots,n;\ j = 1,2,\cdots,m$. This equation can in turn be written as a linear equation that can be solved using standard linear algebra techniques as follows.

Let $M_j = [M_{0,j}\ \ M_{1,j}\ \ M_{2,j}\ \ \cdots\ \ M_{N,j}]^T$ for a DTMC with $N+1$ state space numbered $\{0,1,2,\cdots,N\}$. Further let $\mathbf{e}_j$ be the $(j+1)^{st}$ column of an identity matrix of size $N+1$, with $\mathbf{1}$ being a column vector of ones, then we can write

$$P((I - \mathbf{e}_j\mathbf{e}_j^T) - I)M_j = -\mathbf{1}.$$

Further define $\tilde{P}_j = P((I - \mathbf{e}_j\mathbf{e}_j^T) - I)$, $\tilde{P} = \begin{bmatrix} \tilde{P}_0 & & & \\ & \tilde{P}_1 & & \\ & & \ddots & \\ & & & \tilde{P}_N \end{bmatrix}$, and $VecM = \begin{bmatrix} M_0 \\ M_1 \\ \vdots \\ M_N \end{bmatrix}$,

then we have the linear equation

$$VecM = -\tilde{P}^{-1}\mathbf{1}_{(N+1)^2}. \tag{2.10}$$

Note that $\mathbf{1}_S$ represents a column vector of ones of dimension S.

2.5.3 Example

Consider a DTMC with the state space $\{0,1,2\}$ and transition matrix

$$P = \begin{bmatrix} 0.3 & 0.6 & 0.1 \\ 0.2 & 0.5 & 0.3 \\ 0.6 & 0.2 & 0.2 \end{bmatrix}.$$

We get $VecM = \begin{bmatrix} 3.1471 \\ 3.2353 \\ 2.0588 \\ 1.8000 \\ 2.1400 \\ 2.6000 \\ 4.7826 \\ 3.9130 \\ 4.6522 \end{bmatrix}$, which gives

$$M = \begin{bmatrix} 3.1471 & 1.8000 & 4.7826 \\ 3.2353 & 2.1400 & 3.9130 \\ 2.0588 & 2.6000 & 4.6522 \end{bmatrix}.$$

2.5.4 Mean first recurrence time

Also, M_{ii} is known as the *mean first recurrence time* for state i.

- For every recurrent state i, $f_{ii} = 1$,
- if $M_{ii} < \infty$, we say state i is positive recurrent.
- if $M_{ii} = \infty$, we say state i is null recurrent.

2.6 Absorbing Markov Chains

A Markov chain which has at least one absorbing state is called an absorbing Markov chain. In an absorbing finite Markov chain, every state is either an absorbing state or a transient state. For example, let the states of an absorbing Markov chain be $\mathscr{N} = \{S_1, S_2, ..., S_N\}$, with $|\mathscr{N}| = N$. Let there be T transient states given as $\mathscr{T} = \{S_1, S_2, ..., S_T\}$, with $|\mathscr{T}| = T$ and $\boldsymbol{N - T}$ absorbing states in this chain given as $\mathscr{A} = \{S_{T+1}, S_{T+2}, ..., S_N\}$, where $\mathscr{N} = \mathscr{T} \cup \mathscr{A}$ and $\mathscr{T} \cap \mathscr{A} = 0$. Then the transition matrix can be written as

$$P = \begin{bmatrix} Q & H \\ 0 & I \end{bmatrix},$$

where Q is a square matrix of dimension T, I is a identity matrix of dimension $N - T$ and H is of dimension $T \times (N - T)$. The matrix Q represents the transitions within the transient states and the matrix H represents the transitions from the transient states into the absorbing states. The matrix Q is substochastic, i.e. the sums of each row is less than or equal to 1, but most important is that at least the sum of one of the rows of Q is strictly less than 1.

As an example consider a 5 state Markov chain with state space $\{0, 1, 2, 3, 4\}$ of which two states $\{3, 4\}$ are absorbing states. It will have a transition probability matrix of the form

$$P = \begin{bmatrix} q_{00} & q_{01} & q_{02} & h_{03} & h_{04} \\ q_{10} & q_{11} & q_{12} & h_{13} & h_{14} \\ q_{20} & q_{21} & q_{22} & h_{23} & h_{24} \\ 0 & 0 & 0 & 1 & 0 \\ 0 & 0 & 0 & 0 & 1 \end{bmatrix}.$$

An absorbing Markov chain with only one absorbing state is very common and very useful in representing some distributions encountered in real life, e.g. the phase type distributions. This will be discussed at length in Chapter 3.

2.6.1 Characteristics of an absorbing Markov chain

1. **State of absorption**: It is clear that for an absorbing Markov chain, the system eventually gets absorbed, i.e. the Markov chain terminates eventually. One aspect of interest is usually trying to know into which state the system gets absorbed. Let us define a $T \times (N-T)$ matrix B whose elements b_{ij} represent the probability that the system eventually gets absorbed in state $j \in \mathscr{A}$, given that the system started in the transient state $i \in T$. To obtain the equation for B we first obtain the following. The probability that the system is absorbed into state j at the nth transition, given that it started from a transient state i is given by $\sum_{v=1}^{T}(Q^{n-1})_{iv}H_{vj}$, where $(Q^{n-1})_{iv}$ is the $(i,\ v)$ element of Q^{n-1} and H_{vj} is the $(v,\ j)$ element of H. The argument leading to this is that the system remains in the transient states for the first $n-1$ transitions and then enters the absorbing state j at the nth transition. Hence, b_{ij} is obtained as the infinite sum of the series given by

$$b_{ij} = \sum_{n=1}^{\infty}\sum_{v}^{T}(Q^{n-1})_{iv}H_{vj}.$$

This in matrix form is

$$B = (I + Q + Q^2 + Q^3 + ...)H$$

It is straightforward to show that

$$B = (I - Q)^{-1}H \tag{2.11}$$

provided the inverse exists. This inverse is known to exist if Q is strictly sub-stochastic and irreducible (see Bhat [21]). In our case, Q has to be sub-stochastic. The matrix Q is said to be sub-stochastic if at least for one of the rows i we have $\sum_j Q_{ij} < 1,\ \forall i \in \mathscr{T}$ where Q_{ij} is the (i,j)th element of the matrix Q. The only other requirement for the inverse to exist is that it is irreducible, which will depend on other properties of the specific Markov chain being dealt with.

If we let a_i represent a probability of starting from the transient state $i \in \mathscr{T}$ initially, with vector $\boldsymbol{a} = [a_1, a_2, ..., a_T]$, then the probability of getting absorbed into the absorbing state $j' \in \mathscr{A}$ is given by $b_{j'}$ which is obtained from the vector $\boldsymbol{b} = [b_{1'}, b_{2'}, ..., b_{(N-T)'}]$

$$\boldsymbol{b} = \boldsymbol{a}(I-Q)^{-1}H \tag{2.12}$$

where $j' = T + j$.

2. **Time to absorption:** Let us define a T x $(N-T)$ matrix $C^{(n)}$ whose elements $c^{(n)}_{ij'}$ represent the probability of being absorbed into state $j' \in \mathscr{A}$ at the nth transition, given that the system was in state $i \in \mathscr{T}$ at the start. Then $c^{(n)}_{ij'} = (Q^{n-1})_{iv}H_{vj'}$ and in matrix form $C^{(n)}$ is given as

$$C^{(n)} = Q^{n-1}H, \; n \geq 1$$

If we let the vector $\boldsymbol{c}^{(n)} = [c^{(n)}_{1'}, c^{(n)}_{2'}, ..., c^{(n)}_{(N-T)'}]$, where $c^{(n)}_{j'}$ is the probability that the system gets absorbed into state $j' \in A$ at the nth transition then

$$\boldsymbol{c}^{(n)} = \boldsymbol{a}Q^{n-1}H, \; n \geq 1 \tag{2.13}$$

Of particular interest is the case of an absorbing Markov chain with only one absorbing state. In that case, R is a column vector and $(I-Q)\boldsymbol{e} = H$, where $\boldsymbol{1}$ is a column vector of 1's. Hence,

$$B = \boldsymbol{1}, \;\; C^{(n)} = Q^{n-1}(I-Q)\boldsymbol{1} \text{ and } D = (I-Q)^{-1}\boldsymbol{1}$$

Later we will find that for the case when $N-T = 1$, $c^{(n)}$, $n \geq 1$, is a proper probability mass function. This $C^{(n)}$ is popularly known as the phase type distribution and represented by $(\boldsymbol{a}, Q)$.

3. **Mean time to absorption**: Let us define a $T \times (N-T)$ matrix $\hat{D}$ whose elements $\hat{d}_{ij}$ represent the mean time to absorption into state $j \in A$, given that the system started in state $i \in \mathscr{T}$. Then,

$$\hat{D} = (I + 2Q + 3Q^2 + 4Q^3 + ...)H.$$

After some algebraic manipulations we obtain

$$\hat{D} = (I-Q)^{-2}H. \tag{2.14}$$

Also, if we let the vector $\hat{\boldsymbol{d}} = [\hat{d}_{1'}, \hat{d}_{2'}, ..., \hat{d}_{(N-T)'}]$, where $\hat{d}_{ij'}$ is the mean time to absorption to state j' then

$$\hat{\boldsymbol{d}} = \boldsymbol{a}(I-Q)^{-2}H \tag{2.15}$$

2.6.1.1 Example:

Consider the 5 state Markov chain with state space $\{0,1,2,3,4\}$ of which two states $\{3,4\}$ are absorbing states. Let its transition probability matrix be of the form

$$P = \begin{bmatrix} 0.1 & 0.2 & 0.1 & 0.4 & 0.2 \\ 0.3 & 0.1 & 0.5 & 0.1 & 0.0 \\ 0.4 & 0.0 & 0.25 & 0.3 & 0.05 \\ 0 & 0 & 0 & 1 & 0 \\ 0 & 0 & 0 & 0 & 1 \end{bmatrix}.$$

Here we have

$$I-Q = \begin{bmatrix} 0.9 & -0.2 & -0.1 \\ -0.3 & 0.9 & -0.5 \\ -0.4 & 0.0 & 0.75 \end{bmatrix}, \quad H = \begin{bmatrix} 0.4 & 0.2 \\ 0.1 & 0.0 \\ 0.3 & 0.05 \end{bmatrix}.$$

Suppose we have $\mathbf{a} = [0.1\ 0.6\ 0.3]$, then we have

$$\mathbf{b} = \mathbf{a}(I-Q)^{-1}R = [0.7683\ 0.2317],$$

$$\mathbf{c}^{(5)} = \mathbf{a}Q^4H = [0.0464\ \ 0.0159],$$

$$\hat{\mathbf{d}} = \mathbf{a}(I-Q)^{-2}H = [2.1393\ \ 0.7024].$$

Consider another example with the state space $\{0,1,2,3\}$ and transition matrix

$$P = \begin{bmatrix} 0.1 & 0.2 & 0.1 & 0.6 \\ 0.3 & 0.1 & 0.5 & 0.1 \\ 0.4 & 0.0 & 0.25 & 0.35 \\ 0.0 & 0.0 & 0.0 & 1.0 \end{bmatrix}.$$

Here $\mathbf{b} = 1.0$. If we start with $\mathbf{a} = [0.1\ 0.6\ 0.3]$, then we obtain $\hat{\mathbf{d}} = 2.8417$ and $c^{(n)}$ is obtained as

n	1	2	3	4	5	6	7	8	$\cdots$
$c^{(n)}$	0.2250	0.3287	0.1909	0.1044	0.0623	0.0367	0.0215	0.0126	$\cdots$

2.7 Transient Analysis

If we are interested in transient analysis of a Markov chain, i.e. $Pr\{X_n = j|X_0 = i\}$ we may proceed by any of the following three methods (which are algorithmically only feasible for finite state DTMC), among several other methods. It is known that $\boldsymbol{x}^{(n)} = \boldsymbol{x}^{(n-1)}P = \boldsymbol{x}^{(0)}P^n$.

2.7.1 Direct Algebraic Operations

2.7.1.1 Naive repeated application of P

In this case we simply start with the known values of $\boldsymbol{x}^{(0)}$ and repeatedly apply the relationship

$$\boldsymbol{x}^{(n)} = \boldsymbol{x}^{(n-1)}P,$$

starting from $n = 1$ until the desired time point. For example for a three state DTMC with transition matrix

$$P = \begin{bmatrix} 1/4 & 3/4 & 0 \\ 0 & 1/2 & 1/2 \\ 1/3 & 0 & 2/3 \end{bmatrix},$$

if we are given $\mathbf{x}^{(0)} = [1\ 0\ 0]$ and we want to find the probability distribution of the state of the DTMC at the 4th time epoch we simply apply the recursion four times as

$$\mathbf{x}^{(1)} = \mathbf{x}^{(0)}P,\ \mathbf{x}^{(2)} = \mathbf{x}^{(1)}P,\ \mathbf{x}^{(3)} = \mathbf{x}^{(2)}P,\ \mathbf{x}^{(4)} = \mathbf{x}^{(3)}P,$$

and obtain

$$\mathbf{x}^{(4)} = [0.2122\ \ 0.2695\ \ 0.5182].$$

We will notice that as n gets very large then $\mathbf{x}^{(n+1)} \approx \mathbf{x}^{(n)}$, and in fact the differences get smaller as n increases. This is the limiting behaviour of an ergodic Markov chain and will be presented in detail later.

2.7.1.2 Matrix decomposition approach

A second approach is from observing that the matrix P can be decomposed into the form

$$P = \Lambda D \Lambda^{-1},$$

where D is a diagonal matrix consisting of the eigenvalues of P and Λ is a matrix for which each of its row is the eigenvector corresponding to one of the eigenvalues. Hence we have

$$P^n = \Lambda D^n \Lambda^{-1},$$

which leads to

$$\boldsymbol{x}^{(n)} = \boldsymbol{x}^{(0)} \Lambda D^n \Lambda^{-1}. \tag{2.16}$$

For the example given above we have

$$D = \begin{bmatrix} 0.2083+0.4061i & 0 & 0 \\ 0 & 0.2083-0.4061i & 0 \\ 0 & 0 & 1.0000 \end{bmatrix},$$

$$\Lambda = \begin{bmatrix} 0.7924 & 0.7924 & -0.5774 \\ -0.0440+0.4291i & -0.0440-0.4291i & -0.5774 \\ -0.3228-0.2861i & -0.3228+0.2861i & -0.5774 \end{bmatrix},$$

where the symbol i refers to complex component of the number. This leads to

$$\mathbf{x}^{(4)} = \mathbf{x}^{(0)} \Lambda D^4 \Lambda^{-1} = [0.2122 \;\; 0.2695 \;\; 0.5182].$$

2.7.2 Transient Analysis Based on the z-Transform

Consider the relationship $\boldsymbol{x}^{(n)} = \boldsymbol{x}^{(n-1)} P$ and take the z-transform or the probability generating functions of both sides and let $X(z) = \sum_{n=0}^{\infty} \boldsymbol{x}^{(n)} z^n$, $|z| \leq 1$ we obtain

$$X(z) - x^{(0)} = zX(z)P \Rightarrow X(z) = x^{(0)}[I - zP]^{-1}.$$

We know that $A^{-1} = \frac{1}{\det(A)} \mathrm{Adj}(A)$. While this method is based on a classical theoretical approach, it is not simple to use analytically for a DTMC that has more than 2 states. We use an example to show how this method works. Consider the case of $P = \begin{bmatrix} 1/3 & 3/4 \\ 1/2 & 1/2 \end{bmatrix}$. For this example, we have $\det[I - zP]^{-1} = (1-z)\left(1+\frac{1}{4}z\right)$ and $[I - zP]^{-1} = \frac{1}{1-z}\left(1+\frac{1}{4}z\right)\begin{bmatrix} 1-\frac{1}{2}z & \frac{3}{4}z \\ \frac{1}{2}z & 1-\frac{1}{4}z \end{bmatrix}$. Using partial fractions, we have

$$[I - zP]^{-1} = \frac{1}{1-z}\begin{bmatrix} \frac{2}{5} & \frac{3}{4} \\ \frac{2}{5} & \frac{3}{5} \end{bmatrix} + \frac{1}{1+\frac{1}{4}z}\begin{bmatrix} \frac{3}{5} & -\frac{3}{5} \\ -\frac{2}{5} & \frac{2}{5} \end{bmatrix}$$

Keeping in mind that the inverse of the z-transform $\frac{1}{1-z}$ is the unit step 1 and that of $\frac{1}{1-az}$ is a^n, then we have

$$P^n = \begin{bmatrix} \frac{2}{5} & \frac{3}{5} \\ \frac{2}{5} & \frac{3}{5} \end{bmatrix} + \left(-\frac{1}{4}\right)^n \begin{bmatrix} \frac{3}{5} & -\frac{3}{5} \\ -\frac{2}{5} & \frac{2}{5} \end{bmatrix}$$

This is the transient solution to this chain. We notice that as $n \to \infty$, the second term on the right hand side vanishes and the first term is the limiting or steady state portion of the solution. In general, we can write the closed form of $X(z)$ as

$$X(z) = \frac{1}{1-z}\boldsymbol{x} + \boldsymbol{x}^{(0)} T^g(z)$$

where $T^{g(z)}$ is the transform part that gives the additional part of the solution to lead to transient solution. When it is removed, we only obtain the steady state solution. For this example

$$T^g(z) = \frac{1}{1+\frac{1}{4}z}\begin{bmatrix} \frac{3}{5} & -\frac{3}{5} \\ -\frac{2}{5} & \frac{2}{5} \end{bmatrix}.$$

For details on this method the reader is referred to Howard [56]

2.8 Limiting Behaviour of Markov Chains

One very important key behaviour of interest to us with regards to a time-homogeneous DTMC is its limiting behaviour. i.e. after a long period of time. In studying this limiting behaviour of Markov chains, we need to make clear under what conditions we are able to be specific about this behaviour. We identify the difference between the long term behaviour of an ergodic and a non-ergodic DTMC.

2.8.1 Ergodic Chains

Ergordic Markov chains are those chains that are irreducible, aperiodic and positive recurrent. Proving ergodicity could be very involved in some cases. Let P be the transition matrix of an ergodic DTMC. If the probability vector of the initial state of the system is given by $x^{(0)}$, then the probability vector of the state of the system at any time is given by

$$\boldsymbol{x}^{(n)} = \boldsymbol{x}^{(n-1)}P = \boldsymbol{x}^{(0)}P^n$$

As $n \rightarrow \infty$ we have $\boldsymbol{x}^{(n)} \rightarrow \boldsymbol{x} \implies \boldsymbol{x}^{(n)} \rightarrow \boldsymbol{x}P$, which is known as the invariant (stationary) probability vector of the system.

Similarly, we have

$$P^n \rightarrow \begin{bmatrix} \mathbf{y} \\ \mathbf{y} \\ \vdots \\ \mathbf{y} \end{bmatrix} \quad \text{as } n \rightarrow \infty$$

where $\mathbf{y}$ is known as the limiting or equilibrium or steady state probability vector of the system.

In the case of the ergodic system, we have

$$\boldsymbol{x} = \mathbf{y}.$$

We return to the Example 2.2.1, where we have

$$P = \begin{bmatrix} 0.70 & 0.30 & & & \\ 0.42 & 0.46 & 0.12 & & \\ & 0.42 & 0.46 & 0.12 & \\ & & 0.42 & 0.46 & 0.12 \\ & & & 0.42 & 0.58 \end{bmatrix}$$

Here $\boldsymbol{x}^{(n+1)} = \boldsymbol{x}^{(n)} = \boldsymbol{x}|_{n\to\infty} = [0.5017, 0.3583, 0.1024, 0.0293, 0.0084]$. Similarly if we find that $P^n \to \begin{bmatrix} \mathbf{y} \\ \mathbf{y} \\ \vdots \\ \mathbf{y} \end{bmatrix}$ as $n \to \infty$, where $\mathbf{y} = [0.5017, 0.3583, 0.1024, 0.0293, 0.0084]$.

2.8.2 Non-Ergodic Chains

For non-ergodic systems

$$\boldsymbol{x} \text{ is not necessarily equal to } \mathbf{y}$$

even when both do exist.

For example, consider the transition matrix of the non-ergodic system given by $P = \begin{bmatrix} 0 & 1 \\ 1 & 0 \end{bmatrix}$. This is a periodic system. For this system, $\boldsymbol{x} = [0.5,\ 0.5]$ and $\boldsymbol{y}$ does not exist because the system alternates between the two states. This DTMC is periodic. At any instant in time, $\boldsymbol{y}$ could be $[1,\ 0]$ or $[0,\ 1]$, i.e. we have

$$P^n = \begin{cases} \begin{bmatrix} 0 & 1 \\ 1 & 0 \end{bmatrix}, & \text{if } n \text{ is even} \\ \begin{bmatrix} 1 & 0 \\ 0 & 1 \end{bmatrix}, & \text{if } n \text{ is odd} \end{cases}$$

We consider another example of a non-ergodic system by virtue of not being irreducible. For a reducible DTMC described by a transition matrix of the form

$$P = \begin{bmatrix} 0.5 & 0.5 & & \\ 0.6 & 0.4 & & \\ & & 0.2 & 0.8 \\ & & 0.9 & 0.1 \end{bmatrix}.$$

For this system, the solution to $\mathbf{x} = \mathbf{x}P,\ \ \mathbf{x1} = 1$ does not exist, yet we have $P^n = \begin{bmatrix} \mathbf{y} \\ \mathbf{y} \\ \mathbf{z} \\ \mathbf{z} \end{bmatrix}$, where

$$\mathbf{y} = [0.5445\ \ 0.4545\ \ 0\ \ 0] \neq \mathbf{z} = [0.0\ \ 0.0\ \ 0.5294\ \ 0.4706].$$

Finally, we consider another example of a non-ergodic DTMC by virtue of not being positive recurrent. Suppose we have an infinite number of states $\{0,1,2,3,4,\cdots\}$ and the transition matrix of this is given as

$$P = \begin{bmatrix} 0.3 & 0.7 & & & & \\ 0.1 & 0.1 & 0.8 & & & \\ & 0.1 & 0.1 & 0.8 & & \\ & & 0.1 & 0.1 & 0.8 & \\ & & & \ddots & \ddots & \ddots \end{bmatrix}.$$

This DTMC will neither have an invariant vector nor a steady state vector.

In the rest of the book we focus mainly on ergodic DTMCs.

2.8.3 Mean First Recurrence Time and Steady State Distributions

For an ergodic system we know that x_i is the probability of being in state i at any time, after steady state has been reached. But we also know that M_{ii} is the mean recurrence time for state i. Hence, the probability of being in state i is given as $1/M_{ii}$.

2.9 Numerical Computations of the Invariant Vectors

In this section, we distinguish between finite and infinite state Markov chains as we present the methods. Our interest is to compute the invariant probability vector $\boldsymbol{x}$ rather than the limiting probability vector $\mathbf{y}$. We know and show later that $\boldsymbol{x}$ exists for every irreducible positive recurrent Markov chain. We therefore focus on such chains only.

We want to determine $\boldsymbol{x}$ such that

$$\boldsymbol{x} = \boldsymbol{x}P \text{ and } \boldsymbol{x}\mathbf{1} = 1. \tag{2.17}$$

2.9.1 Finite State Markov Chains

There are mainly two types of methods used for computing the stationary vector of a DTMC; direct methods and iterative methods. Iterative methods are used more frequently for most DTMC which have a large state space – very common in most practical situations. However, the computational effort for iterative methods are dependent on the size and structure of the DTMC and also the starting solution. Direct methods are suitable for smaller DTMC and the number of steps required to get a solution is known ahead of time.

For the iterative methods we work with the equation in the form

$$\boldsymbol{x} = \boldsymbol{x}P, \ \boldsymbol{x}\mathbf{1} = 1,$$

whereas for most direct methods we convert the problem to a classical linear algebra problem of the form $AX = b$, by creating a matrix $\tilde{P}$ which is the same as $I - P$, with the last column replaced by a column of ones. Our problem (of N states Markov chain) then becomes

$$\mathbf{e}_N^T = \boldsymbol{x}\tilde{P} \ \rightarrow \ \tilde{P}^T \boldsymbol{x}^T = \mathbf{e}_N,$$

which is now in the standard form. We use the notation $\mathbf{e}_j$ to represent a column vector which has all zero entries except in the jth column that we have 1 and $\mathbf{e}_j^T$ is its transpose.

For example, consider a three state DTMC $\{0, 1, 2\}$ with transtion matrix P given as

$$P = \begin{bmatrix} 0.3 & 0.6 & 0.1 \\ 0.2 & 0.3 & 0.5 \\ 0.4 & 0.5 & 0.1 \end{bmatrix}.$$

We will use this example in all of our numerical computation methods for invariant vectors.

For the iterative method we work with

$$[x_0 \ x_1 \ x_2] = [x_0 \ x_1 \ x_2] \begin{bmatrix} .3 & .6 & .1 \\ .2 & .3 & .5 \\ .4 & .5 & .1 \end{bmatrix}, \ [x_1 \ x_2 \ x_3] \begin{bmatrix} 1.0 \\ 1.0 \\ 1.0 \end{bmatrix} = 1,$$

whereas for the direct method we work with

$$\begin{bmatrix} 0.7 & -0.2 & -0.4 \\ -0.6 & 0.7 & -0.5 \\ 1.0 & 1.0 & 1.0 \end{bmatrix} \begin{bmatrix} x_1 \\ x_2 \\ x_3 \end{bmatrix} = \begin{bmatrix} 0.0 \\ 0.0 \\ 1.0 \end{bmatrix},$$

where

$$\tilde{P} = \begin{bmatrix} 0.7 & -0.6 & 1.0 \\ -0.2 & 0.7 & 1.0 \\ -0.4 & -0.5 & 1.0 \end{bmatrix} \rightarrow \tilde{P}^T = \begin{bmatrix} .7 & -.2 & -.4 \\ -.6 & .7 & -.5 \\ 1 & 1 & 1 \end{bmatrix}.$$

Direct methods do not involve iterations. They simply involve a set of operations which is usually of a known fixed number, depending on the size of the Markov chain. They are efficient for small size Markov chains and usually very time consuming for large Markov chains. One direct method which seems to be able to handle a large size Markov chain is the state reduction method (by Grassmann, Taksar Heyman [46]). Theoretically, one may state that it has no round off errors. The most popular of the direct techniques is the Gaussian elimination. For more details of this algorithm as applicable to Markov chains, see Stewart [93].

2.9.1.1 Direct Methods

In what follows we present some of the well known direct methods. We present only a synoptic kind of view of this approach in this book. First we assume that the equations have been transformed to the standard form

$$AX = b. \tag{2.18}$$

Using the above example we have

$$P = \begin{bmatrix} 0.3 & 0.6 & 0.1 \\ 0.2 & 0.3 & 0.5 \\ 0.4 & 0.5 & 0.1 \end{bmatrix},$$

which results in

$$A = \begin{bmatrix} 0.7 & -0.2 & -0.4 \\ -0.6 & 0.7 & -0.5 \\ 1.0 & 1.0 & 1.0 \end{bmatrix}, X = \begin{bmatrix} x_1 \\ x_2 \\ x_3 \end{bmatrix} b = \begin{bmatrix} 0.0 \\ 0.0 \\ 1.0 \end{bmatrix}.$$

- **Inverse matrix approach**: This is the simplest of all the direct methods. It simply solves for X as $X = A^{-1}b$, assuming that the inverse of A exists. For most of the finite DTMC we will be discussing in this book, the inverse of A would exist. From the above example we have

$$\boldsymbol{x}^T = A^{-1}b = \begin{bmatrix} 0.8955 & -0.1493 & 0.2836 \\ 0.0746 & 0.8209 & 0.4403 \\ -0.9701 & -0.6716 & 0.2761 \end{bmatrix} \begin{bmatrix} 0.0 \\ 0.0 \\ 1.0 \end{bmatrix} = \begin{bmatrix} 0.2836 \\ 0.4403 \\ 0.2761 \end{bmatrix}.$$

- **Gaussian Elimination**: This is another popular method which is more suitable for moderate size matrices than the inverse approach. The matrix A is transformed through row operations into an upper triangular matrix U, such that $U\boldsymbol{x}^T = b = \mathbf{e}_N$ and then the elements of $\boldsymbol{x}$ are obtained from back substitution as

$$x_N = 1/u_{N,N}, \ x_i = -(\sum_{k=i+1}^{N} u_{i,k}x_k)/u_{i,i}, i = 0, 1, 2, \cdots, N-1. \tag{2.19}$$

For this example, we have

$$U = \begin{bmatrix} 0.7 & -0.2 & -0.4 \\ & 0.5286 & -0.8429 \\ & & 3.6216 \end{bmatrix}.$$

Applying the back substitution equation we obtain

$$\boldsymbol{x}^T = \begin{bmatrix} 0.2836 \\ 0.4403 \\ 0.2761 \end{bmatrix}.$$

- **State Reduction Method**:This method is also popularly known as the GTH (Grassmann, Taksar and Heyman [46]) method. It is actually based on Gaussian elimination with state space reduction during each iteration and it uses the form of equation $\boldsymbol{x} = \boldsymbol{x}P$, $\boldsymbol{x}\mathbf{1} = 1$, unlike the other direct methods. Consider the equation $x_j = \sum_{i=0}^{N} x_i p_{ij}$, $j = 0, 1, 2, ..., N$.
  The procedure starts by eliminating x_N from the Nth equation to obtain

$$x_N = \sum_{i=0}^{N-1} x_i p_{iN}/(1 - p_{N,N}).$$

Replacing $1 - p_{NN}$ with $\sum_{j=0}^{N-1} p_{N,j}$ and using the property that $x_j = \sum_{i=0}^{N-1} x_i p_{ij}^{(N-1)}$, $j = 0, 1, 2, ..., N-1$, where

$$p_{i,j}^{(N-1)} = p_{i,j} + p_{i,N} p_{N,j}/(1 - p_{N,N}) = p_{i,j} + p_{i,N} p_{N,j}/\sum_{j=0}^{N-1} p_{N,j}.$$

This is the principle by which state reduction works.
Generally, $x_n = \sum_{i=1}^{n-1} x_i q_{i,n}$, $n = N, N-1, N-2, ..., 2, 1$ where $q_{i,n} = p_{i,n}^{(n)}/\sum_{j=0}^{n-1} p_{n,j}^{(n)}$.
The algorithm can be stated as follows (Grassmann, Taksar and Heyman [46]) referred to as the GTH:

GTH Algorithm:

1. For $n = N, N-1, N-2, ..., 2, 1$, do

$$q_{i,n} \leftarrow p_{i,n}/\sum_{j=0}^{n-1} p_{n,j}, \quad i = 0, 1, ..., n-1$$

$$p_{i,j} \leftarrow p_{i,j} + q_{i,n} p_{n,j}, \quad i, j = 0, 1, ..., n-1$$

$$r_1 \leftarrow 1$$

2. For $j = 2, 3, ..., N$, $r_j \leftarrow \sum_{i=0}^{j-1} r_i q_{i,j}$
3. For $j = 2, 3, ..., N$, $x_j \leftarrow r_j/\sum_{i=0}^{N} r_i$, $j = 0, 1, ..., N$

The advantage with this approach is that it is numerically stable even if the system is ill-conditioned.

2.9.1.2 Iterative Methods:

Iterative methods are usually more efficient and more effective than the direct methods when the size of the matrix is large. The number of iterations required to achieve the desired convergence is not usually known in advance and the resulting distributions have round-off errors associated with them.

Four commonly used iterative methods will be discussed in this section. They are: i)the Power method which is more applicable to transition matrices, ii) the Jacobi method, iii) the Gauss-Seidel method, and iv) the method of successive over-relaxation.

- **The Power Method**: The power method simply involves starting with a probability vector $\boldsymbol{x}^{(0)}$ which satisfies the condition $\boldsymbol{x}^{(0)}\mathbf{1} = 1$ and then applying the relationship $\boldsymbol{x}^{(n+1)} = \boldsymbol{x}^{(n)}P$, $n \geq 0$ until $[\boldsymbol{x}^{(n+1)} - \boldsymbol{x}^{(n)}]_i < \varepsilon$, $\forall i$, then we stop the iteration, where ε is the convergence criterion. Often the number of iterations required depends on both P and the starting vector $\boldsymbol{x}^{(0)}$. It is a very naive approach which is only useful for small size problems, and especially if we also need to know the transient behaviour of the DTMC before it reaches steady state. The next three approaches will be discussed here in a modified form for the transition matrix.
First let us write $Q = P - I$. Our interest is to solve for $\boldsymbol{x} = [x_1, x_2, ..., x_n]$, where $\boldsymbol{x}Q = 0$ and $\boldsymbol{x}\mathbf{1} = 1$. In order to conform to the standard representations of these methods we write the same equation in a transposed form as $Q^T\boldsymbol{x}^T = 0$.
Let us now separate Q^T into a diagonal component D, an upper triangular component U and a lower triangular component L with ij elements given as $-d_{ii}$, u_{ij} and l_{ij}, respectively. It is assumed that $d_{ii} > 0$. Note that the D, L and U are not the same as the well known LU or LDU decompositions. For example given a matrix

$$Q^T = \begin{bmatrix} q_{0,0} & q_{1,0} & q_{2,0} \\ q_{0,1} & q_{1,1} & q_{2,1} \\ q_{0,2} & q_{1,2} & q_{2,2} \end{bmatrix},$$

we can write $Q^T = L - D + U$, where

$$L = \begin{bmatrix} 0 & 0 & 0 \\ q_{0,1} & 0 & 0 \\ q_{0,2} & q_{1,2} & 0 \end{bmatrix}, \quad U = \begin{bmatrix} 0 & q_{1,0} & q_{2,0} \\ 0 & 0 & q_{2,1} \\ 0 & 0 & 0 \end{bmatrix},$$

$$D = \begin{bmatrix} -q_{0,0} & 0 & 0 \\ 0 & -q_{1,1} & 0 \\ 0 & 0 & -q_{2,2} \end{bmatrix}.$$

- **The Jacobi Method**: Then the Jacobi iteration is given in the scalar form as

$$x_i^{(k+1)} = \frac{1}{d_{ii}}\{\sum_{j \neq i}(l_{ij} + u_{ij})x_j^{(k)}\}, \quad i = 0, 1, 2, ..., N \tag{2.20}$$

or in matrix form, it can be written as

$$(\boldsymbol{x}^{(k+1)})^T = D^{-1}(L+U)(\boldsymbol{x}^{(k)})^T. \tag{2.21}$$

For example, consider the Markov chain given by

$$P = \begin{bmatrix} p_{0,0} & p_{0,1} & p_{0,2} \\ p_{1,0} & p_{1,1} & p_{1,2} \\ p_{2,0} & p_{2,1} & p_{2,2} \end{bmatrix} = \begin{bmatrix} 0.1 & 0.2 & 0.7 \\ 0.3 & 0.1 & 0.6 \\ 0.2 & 0.5 & 0.3 \end{bmatrix}$$

Suppose we have an intermediate solution $\boldsymbol{x}^{(k)}$ for which $\boldsymbol{x}^{(k)}\mathbf{1} = 1$, then we have

$$x_0^{(k+1)} = \frac{1}{1-p_{0,0}}[x_1^{(k)}p_{1,0} + x_2^{(k)}p_{2,0}]$$

$$x_1^{(k+1)} = \frac{1}{1-p_{1,1}}[x_0^{(k)}p_{0,1} + x_2^{(k)}p_{2,1}]$$

$$x_2^{(k+1)} = \frac{1}{1-p_{2,2}}[x_0^{(k)}p_{0,2} + x_1^{(k)}p_{1,2}]$$

- **Gauss-Seidel Method**: The Gauss-Seidel method uses the most up-to-date information on solutions available for computations. It is given by the following iteration in the scalar form:

$$x_i^{(k+1)} = \frac{1}{d_{ii}}\{\sum_{j=0}^{i-1} l_{ij}x_j^{(k+1)} + \sum_{j=i+1}^{n} u_{ij}x_j^{(k)}\} \tag{2.22}$$

and in matrix form as

$$(\boldsymbol{x}^{(k+1)})^T = D^{-1}(L(\boldsymbol{x}^{(k+1)})^T + U(\boldsymbol{x}^{(k)})^T) \tag{2.23}$$

This method seems to be more frequently used than the Jacobi. Using the above example, we have

$$x_0^{(k+1)} = \frac{1}{1-p_{0,0}}[x_1^{(k)}p_{1,0} + x_2^{(k)}p_{2,0}]$$

$$x_1^{(k+1)} = \frac{1}{1-p_{1,1}}[x_0^{(k+1)}p_{0,1} + x_2^{(k)}p_{2,1}]$$

$$x_2^{(k+1)} = \frac{1}{1-p_{2,2}}[x_0^{(k+1)}p_{0,2} + x_1^{(k+1)}p_{1,2}]$$

- **The Successive Over-relaxation Method**: This method introduces an extra parameter ω. The method is a generalization of the Gauss-Seidel. The iteration is implemented as follows:

$$x_i^{(k+1)} = (1-\omega)x_i^{(k)} + \omega\{\frac{1}{d_{ii}}\{\sum_{j=0}^{i-1} l_{ij}x_j^{(k+1)} + \sum_{j=i+1}^{N} u_{ij}x_j^{(k)}\}\}, \quad (2.24)$$

and in matrix form

$$(\boldsymbol{x}^{(k+1)})^T = (1-\omega)(\boldsymbol{x}^{(k)})^T + \omega\{D^{-1}(L(\boldsymbol{x}^{(k+1)})^T + U(\boldsymbol{x}^{(k)})^T)\}. \quad (2.25)$$

Using the above example, we have

$$x_0^{(k+1)} = (1-\omega)x_0^{(k)} + \omega\frac{1}{1-p_{0,0}}[x_1^{(k)}p_{1,0} + x_2^{(k)}p_{2,0}]$$

$$x_1^{(k+1)} = (1-\omega)x_1^{(k)} + \omega\frac{1}{1-p_{1,1}}[x_0^{(k+1)}p_{0,1} + x_2^{(k)}p_{2,1}]$$

$$x_2^{(k+1)} = (1-\omega)x_2^{(k)} + \omega\frac{1}{1-p_{2,2}}[x_0^{(k+1)}p_{0,2} + x_1^{(k+1)}p_{1,2}]$$

The value of ω is usually selected between 0 and 2. When $\omega < 1$, we say we have an under-relaxation and when $\omega > 1$, we have the Gauss-Seidel iteration. The greatest difficulty with this method is how to select the value of ω.
The Jacobi method is very appropriate for parallelization of computing. Note that diagonal dominance of the matrix $I-P$ is required for the Jacobi and in fact also for the Gauss-Seidel methods but it is always satisfied for a Markov chain.

2.9.2 Bivariate Discrete Time Markov Chains

Consider a bivariate process $\{(X_n, Y_n), n = 0, 1, 2, \cdots,\}$ that has a Markov structure. We assume that X_n assumes values in the set $\mathscr{I}$ and Y_n assumes values in the set $\mathscr{M} = \{1, 2, \cdots, M < \infty\}$. An example of this is a system that has two classes of data packets of types 1 and 2, where type 1 has a higher priority than type 2. Suppose we do not allow more than M of type 2 packets to be in the system at any one time, we can represent X_n as the number of type 1 and Y_n as the number of type 2 packets in the system at time n. If we assume that $\{(X_n, Y_n); n = 0, 1, 2, \cdots\}$ is a DTMC we can write the transition matrix of this Markov chain as

$$P = \begin{bmatrix} P_{0,0} & P_{0,1} & P_{0,2} & \cdots \\ P_{1,0} & P_{1,1} & P_{1,2} & \cdots \\ P_{2,0} & P_{2,1} & P_{2,2} & \cdots \\ \vdots & \vdots & \vdots & \vdots \end{bmatrix}, \quad (2.26)$$

where $P_{i,j}, (i,j) = 0, 1, 2, \cdots$ is a block matrix of order $M \times M$. The element $(P_{i,j})_{l,k}$, $(i,j) = 0, 1, 2, \cdots; (l,k) = 1, 2, \cdots, M$, is $Pr\{X_{n+1} = j, Y_{n+1} = k | X_n = i, Y_n = l\}$. Usually we call the set $L_i = \{(i,1), (i,2), \cdots, (i,M)\}$ level i. We also say in state (i,l) that we are in level i and phase l.

Define $x_{i,k}^{(n)} = Pr\{X_n = i, Y_n = k\}$ and write $\boldsymbol{x}^{(n)} = [\boldsymbol{x}_0^{(n)}\ \boldsymbol{x}_1^{(n)}\ \boldsymbol{x}_2^{(n)}\ \cdots]$ with $\boldsymbol{x}_i^{(n)} = [x_{i,1}^{(n)}\ x_{i,2}^{(n)}\ \cdots\ x_{i,M}^{(n)}]$. We have

$$\boldsymbol{x}^{(n+1)} = \boldsymbol{x}^{(n)}P.$$

If the DTMC is positive recurrent then we have $x_{i,k}^{(n)}|_{n\to\infty} = x_{i,k}$, hence

$$\boldsymbol{x} = \boldsymbol{x}P, \ \ \boldsymbol{x}\mathbf{1} = 1.$$

2.9.3 *Computing Stationary Distribution for the Finite Bivariate DTMC*

The methods presented in Section 2.6.1 are easily extended to the bivariate and all multivariate finite DTMC.

To avoid unnecessary notational complexities we present only the case where $P_{i,j}$ is of dimension $M \times M$, $\forall (i,j)$. We can easily modify the result for cases where the dimension of $P_{i,j}$ depends on i and j. Let $\mathbf{1}_M$ be an M column vector of ones and $\mathbf{0}_M$ an $M \times 1$ column vector of zeros. Assume X_n has a state space given as $0, 1, 2, \cdots, N < \infty$, then we can write the stationary equation as

$$\begin{array}{ccccccccccc}
\mathbf{0}_M^T & = & \boldsymbol{x}_0(P_{0,0}-I) & + & \boldsymbol{x}_1 P_{1,0} & + & \boldsymbol{x}_2 P_{2,0} & + & \cdots & + & \boldsymbol{x}_N P_{N,0} \\
\mathbf{0}_M^T & = & \boldsymbol{x}_0 P_{0,1} & + & \boldsymbol{x}_1(P_{1,1}-I) & + & \boldsymbol{x}_2 P_{2,1} & + & \cdots & + & \boldsymbol{x}_N P_{N,1} \\
\mathbf{0}_M^T & = & \boldsymbol{x}_0 P_{0,2} & + & \boldsymbol{x}_1 P_{1,2} & + & \boldsymbol{x}_2(P_{2,2}-I) & + & \cdots & + & \boldsymbol{x}_N P_{N,2} \\
\vdots & = & \vdots & + & \vdots & + & \vdots & + & \cdots & + & \vdots \\
\mathbf{0}_M^T & = & \boldsymbol{x}_0 P_{0,N-1} & + & \boldsymbol{x}_1 P_{1,N-1} & + & \cdots & + & \boldsymbol{x}_{N-1}(P_{N-1,N-1}-I) & + & \boldsymbol{x}_N P_{N,N-1} \\
\mathbf{1}_M^T & = & \boldsymbol{x}_0 & + & \boldsymbol{x}_1 & + & \boldsymbol{x}_2 & + & \cdots & + & \boldsymbol{x}_N
\end{array}.$$

After transposing we have

$$\begin{bmatrix} P_{0,0}-I & P_{1,0} & P_{2,0} & \cdots & P_{N,0} \\ P_{0,1} & P_{1,1}-I & P_{2,1} & \cdots & P_{N,1} \\ \vdots & \vdots & \vdots & \cdots & \vdots \\ P_{0,N-1} & P_{1,N-1} & P_{2,N-1} & \ddots & P_{N,N-1} \\ \mathbf{1} & \mathbf{1} & \mathbf{1} & \cdots & \mathbf{1} \end{bmatrix} \begin{bmatrix} \boldsymbol{x}_0 \\ \boldsymbol{x}_1 \\ \vdots \\ \boldsymbol{x}_{N-1} \\ \boldsymbol{x}_N \end{bmatrix} = \begin{bmatrix} \mathbf{0}_M \\ \mathbf{0}_M \\ \vdots \\ \mathbf{0}_M \\ \mathbf{1}_M \end{bmatrix}. \tag{2.27}$$

This is now of the form similar to

$$AX = b$$

and can be solved using the techniques discussed earlier. The techniques are briely listed here:

- **Direct Methods**

- **The inverse matrix approach**: This approach can be used for the bivariate DTMC just like in the case of the univariate DTMC. Simply we compute $\boldsymbol{x}$ from
$$X = A^{-1}b. \tag{2.28}$$
- **The block Gaussian elimination method**: The matrix A is transformed through block row operations into an upper triangular block matrix U, such that $U\boldsymbol{x}^T = b$. The triangular matrix consists of block matrices $U_{i,j}$, for which $U_{i,j} = 0,\ i > j$. The block vectors $(\boldsymbol{x}_i)$ of $\boldsymbol{x}$ are obtained from back substitution as
$$\boldsymbol{x}_N^T = u_{N,N}^{-1}\mathbf{1},\ \ \boldsymbol{x}_i^T = -U_{i,i}^{-1}(\sum_{k=i+1}^{N} U_{i,k}\boldsymbol{x}_k^T), i = 0,1,2,\cdots,N-1. \tag{2.29}$$
- **Block state reduction**: This technique and its scalar counter part are based on censoring of the Markov chain. Consider the DTMC with $N+1$ levels labelled $\{0,1,2,\cdots,N\}$. For simplicity we will call this the block state space $\{0,1,2,\cdots,N\}$ and let it be partitioned into two subsets E and $\bar{E}$ such that $E \cup \bar{E} = \{0,1,2,\cdots,N\}$ and $E \cap \bar{E} = \emptyset$. For example, we may have $E = \{0,1,2,\cdots,M\}$ and $\bar{E} = \{M+1, M+2,\cdots,N\}$, $M \leq N$. The transition matrix of the DTMC can be written as
$$P = \begin{bmatrix} H & S \\ T & U \end{bmatrix}.$$
We can write the stationary vector of this DTMC as $\boldsymbol{x} = [\boldsymbol{x}_e\ \boldsymbol{x}_{e'}]$ from which we obtain
$$\boldsymbol{x}_e = \boldsymbol{x}_e H + \boldsymbol{x}_{e'} T,\ \ \boldsymbol{x}_e' = \boldsymbol{x}_e S + \boldsymbol{x}_{e'} U. \tag{2.30}$$
From this we obtain
$$\boldsymbol{x}_e = \boldsymbol{x}_e(H + S(I-U)^{-1}T). \tag{2.31}$$
The matrix $H + S(I-U)^{-1}T$ is a new transition marix and it is stochastic. Its stationary distribution is the vector $\boldsymbol{x}_e$ which can be obtained as $\boldsymbol{x}_e = \boldsymbol{x}_e P_e,\ \boldsymbol{x}_e\mathbf{1} = 1$ and normalized for it to be stochastic. It is immediately clear that we have reduced or restricted this DTMC to be observed only when it is in the states of E. This is a censored DTMC. This idea is used to block-reduce the DTMC and apply the block state reduction to solve for $\boldsymbol{x}$ in the original DTMC.
In block state reduction we first partition the DTMC into two sets with the first set given as $E = \{0,1,2,\cdots,N-1\}$ and the second set $\bar{E} = \{N\}$. We study the states of E only as censored DTMC, and then partition it into $E = \{0,1,2\cdots,N-2\}$ and $\bar{E} = \{N-1\}$, and the process continues until we reduce it to just the set $E = \{0\}$. We obtain the stationary distribution of this reduced DTMC. We then start to expand it to states $E = \{0,1\}$ and continue to expand until we reach the original set of states $\{0,1,2,\cdots,N\}$.

We demonstrate how this procedure works through a small example. Consider a bivariate DTMC that has four levels in its states given by $\{0,1,2,3\}$ and the phases are of finite dimension M. Its transition matrix P has block elements $P_{i,j}$ of order $M \times M$. If we partition the state space into $E = \{0,1,2\}$ and $\bar{E} = \{3\}$, then

$$P_e = \begin{bmatrix} P_{00} & P_{01} & P_{02} \\ P_{1,0} & P_{11} & P_{1,2} \\ P_{2,0} & P_{2,1} & P_{2,2} \end{bmatrix} + \begin{bmatrix} P_{03} \\ P_{13} \\ P_{23} \end{bmatrix} (I - P_{33})^{-1} \begin{bmatrix} P_{30} & P_{31} & P_{32} \end{bmatrix}.$$

When this new set $\{0,1,2\}$ is further partitioned into $E = \{0,1\}$ and $\bar{E} = \{2\}$ we get another transition matrix P_e. Another step further leads to a partitioning into $E = \{0\}$ and $\bar{E} = \{1\}$. Writing $\boldsymbol{x} = \boldsymbol{x}P$, we can write the last block equation of this stationary distribution as

$$\boldsymbol{x}_3 = \boldsymbol{x}_0 P_{0,3} + \boldsymbol{x}_1 P_{1,3} + \boldsymbol{x}_2 P_{2,3} + \boldsymbol{x}_3 P_{33}.$$

From this we have

$$\boldsymbol{x}_3 = (\boldsymbol{x}_0 P_{0,3} + \boldsymbol{x}_1 P_{1,3} + \boldsymbol{x}_2 P_{2,3})(I - P_{33})^{-1}.$$

The block equation before the last can be writen as

$$\boldsymbol{x}_2 = \boldsymbol{x}_0 P_{0,2} + \boldsymbol{x}_1 P_{1,2} + \boldsymbol{x}_2 P_{2,2} + \boldsymbol{x}_3 P_{32},$$

which can be also be written as

$$\boldsymbol{x}_2 = \boldsymbol{x}_0 (P_{0,2} + P_{0,3} K_{3,2}) + \boldsymbol{x}_1 (P_{1,2} + P_{1,3} K_{3,2}) + \boldsymbol{x}_2 (P_{2,2} + P_{2,3} K_{3,2}),$$

where $K_{3,2} = (I - P_{3,3})^{-1} P_{3,2}$. This in turn leads to

$$\boldsymbol{x}_2 = (\boldsymbol{x}_0 (P_{0,2} + P_{0,3} K_{3,2}) + \boldsymbol{x}_1 (P_{1,2} + P_{1,3} K_{3,2}))(I - (P_{2,2} + P_{2,3} K_{3,2}))^{-1}.$$

This process is continued further to reduce the block equation and obtain $\boldsymbol{x}_1$ in terms of $\boldsymbol{x}_0$ and then solve for $\boldsymbol{x}_0$ after one more reduction. This process is then reversed to expand into obtaining $\boldsymbol{x}_1$ from $\boldsymbol{x}_0$, followed by obtaining $\boldsymbol{x}_2$ using $\boldsymbol{x}_0$ and $\boldsymbol{x}_1$ and finally $\boldsymbol{x}_3$ using $\boldsymbol{x}_0$, $\boldsymbol{x}_1$ and $\boldsymbol{x}_2$. The final result is then nomalized. A detailed description of this algorithm together with the steps can be found in [46] .

- **Iterative Methods**: The same sets of methods described earlier can be used for bivariate DTMC, i.e. the power method, Jacobi, Gauss-Seidel and the Successive over relaxation. The power method approach is straightforward and will not be elaborated on here.

  First let us write $Q = P - I$. In that case, our interest is to solve for $\boldsymbol{x} = [\boldsymbol{x}_0\ \boldsymbol{x}_1\ \boldsymbol{x}_2\ \cdots\ \boldsymbol{x}_n]$, where $\boldsymbol{x}Q = 0$ and $\boldsymbol{x}\,\boldsymbol{1} = 1$. In order to conform to the stan-

dard representations of these methods we write the same equation in a transposed form as $Q^T x^T = 0$.
Let us now separate Q^T into a diagonal component D, an upper triangular component U and a lower triangular component L with ij block elements given as $-D_{ij}$, U_{ij} and L_{ij}, respectively. It is assumed that the inverse of D_{ii} exists. Note that the D, L and U are not the same as the well known LU or LDU decompositions.

- **The Jacobi Method**: Then the Jacobi iteration is given in the scalar form as

$$(\boldsymbol{x}^{(k+1)})^T = D^{-1}(L+U)(\boldsymbol{x}^{(k)})^T \tag{2.32}$$

- **Gauss-Seidel Method**: The Gauss-Seidel method uses the most up-to-date information on solutions available for computations. It is given by the following iteration

$$(\boldsymbol{x}^{(k+1)})^T = (D)^{-1}(L(\boldsymbol{x}^{(k+1)})^T + U(\boldsymbol{x}^{(k)})^T). \tag{2.33}$$

- **The Successive Over-relaxation Method**: This method introduces an extra parameter ω. The method is a generalization of the Gauss-Seidel. The iteration is implemented as follows:

$$(\boldsymbol{x}^{(k+1)})^T = (1-\omega)(\boldsymbol{x}^{(k)})^T + \omega\{D^{-1}(L(\boldsymbol{x}^{(k+1)})^T + U(\boldsymbol{x}^{(k)})^T)\}. \tag{2.34}$$

2.9.4 Special Finite State DTMC

For special types of finite state DTMC which are of the block tridiagonal structures there are specially effective algorithms for obtaining the stationary vectors. Such a DTMC has a matrix of the form

$$P = \begin{bmatrix} A_0 & U_0 & & & \\ D_1 & A_1 & U_1 & & \\ & \ddots & \ddots & \ddots & \\ & & & D_N & A_N \end{bmatrix}, \tag{2.35}$$

where the entries $A_i, U_i, D_i, i = 0, 1, \cdots, N$ could be block matrices of finite dimensions. Our interest is to solve for the vector $\boldsymbol{x} = [\boldsymbol{x}_0, \boldsymbol{x}_1, \cdots, \boldsymbol{x}_N]$ where

$$\boldsymbol{x} = \boldsymbol{x}P, \ \boldsymbol{x}\mathbf{1} = 1,$$

and $\boldsymbol{x}_k$, $k = 0, 1, 2, \cdots, N$, is also a vector.
Three methods are presented here due to Gaver, Jacobs and Latouche [45], Grassmann, Taksar and Heyman [46], and Ye and Li [102]. All the three methods are based on direct methods relating to the Gaussian elimintation.

- **Level Reduction**: This method is due to Gaver, Jacobs and Latouche [45]. It is based on Gaussian elmination and closely related to the block state reduction which is due to Grassmann, Taksar and Heyman [46]. They are all based on censored Markov chain idea in that a set of states is restricted – thereby the name level reduction. We consider the level reduced DTMC associated with this Markov chain at the kth stage, we can write

$$P_k = \begin{bmatrix} C_k & U_k & & \\ D_{k+1} & A_{k+1} & U_{k+1} & \\ & \ddots & \ddots & \ddots \\ & & D_N & A_N \end{bmatrix}, \quad 0 \le k \le N-1, \tag{2.36}$$

where

$$C_0 = A_0, \; C_k = A_k + D_k(I - C_{k-1})^{-1} U_{k-1}, \; 1 \le k \le N.$$

The solution is now given as

$$\boldsymbol{x}_N = \boldsymbol{x}_N C_N, \tag{2.37}$$

$$\boldsymbol{x}_k = \boldsymbol{x}_{k+1} D_{k+1} (I - C_k)^{-1}, \; 0 \le k \le N-1, \tag{2.38}$$

and

$$\sum_{k=0}^{N} \boldsymbol{x}_k \mathbf{1} = 1.$$

The first equation is solved using one of the methods proposed earlier, if the blocks of matrices are of dimension more than one.

- **Block State Reduction**: This method which is due to Grassmann, Taksar and Heyman [46], is the reverse of the first method. We censor the Markov chain from the bottom as follows

$$P_{N-k} \begin{bmatrix} A_0 & U_0 & & \\ D_1 & A_1 & U_1 & \\ & \ddots & \ddots & \ddots \\ & & D_{N-k} & E_{N-k} \end{bmatrix}, \quad 0 \le k \le N-1, \tag{2.39}$$

where

$$E_N = A_N, \; E_{N-k} = A_{N-k} + U_{N-k}(I - A_{N-k+1})^{-1} D_{N-k+1}, \; 1 \le k \le N.$$

The solution is now given as

$$\boldsymbol{x}_0 = \boldsymbol{x}_0 E_0, \tag{2.40}$$

$$\boldsymbol{x}_{N-k} = \boldsymbol{x}_{N-k-1} U_{N-k-1} (I - E_{N-k})^{-1}, \; 1 \le k \le N, \tag{2.41}$$

and

$$\sum_{k=0}^{N} \boldsymbol{x}_k \mathbf{1} = 1.$$

- **Folding Algorithm**: This method is related to the first two in that they all group the DTMC into two classes. The folding algorithm groups them into odd and even whereas the other two group the DTMC into one and others. It then rearranges the DTMC as $\{0,2,4,6,\cdots,N,1,3,5,\cdots,N-1\}$, if N is even, as an example. The transition matrix is then partitioned accordingly leaving it in a good structure. For the details about this algorithm you are referred to Ye and Li [102].

2.9.5 Infinite State Markov Chains

Consider a DTMC $\{X_n,\ n=0,1,2,\cdots\}$, $X_n=0,1,2,\cdots$. Since $X_n \in \mathscr{I}$ we say its an infinite DTMC. The finite case is when $X_n \in \{0,1,2,\cdots,N<\infty\}$. If the infinite DTMC is positive recurrent we know that it has a stationary distribution $\boldsymbol{x}$ such that

$$\boldsymbol{x} = \boldsymbol{x}P, \ \ \boldsymbol{x}\mathbf{1} = 1.$$

Even if we know that it has a stationary distribution we still face a challenge on how to compute this distribution since all the methods presented for finite DTMC can no longer be used because the state space is infinite.

One may suggest that we truncate the infinite DTMC at an appropriate state $N<\infty$ to a finite DTMC and apply the earlier techniques. A truncation method could involve selecting the smallest N such that $1-\sum_{j=0}^{N} p_{ij} < \varepsilon,\ \forall i$, where ε is a very small value, say 10^{-12}, for example. Such truncations could lead to problems at times. However for some special infinite DTMCs there are well established methods for determining the stationary distribution without truncation. In addition there are methods for establishing positive recurrency for such DTMCs. We will be presenting those classes of infinite DTMCs.

First we discuss the general case of infinite DTMC and present what is known about them. We then present the three classes of infinite DTMCs, called the GI/M/1 types, M/G/1 types and QBD types, which are numerically tractable and for which the results are well known. Since the results to be developed here are valid for both univariate and multivariate DTMCs, with only the first variable allowed to be unbounded, we will present results for the bivariate case.

2.9.6 Some Results for Infinite State Markov Chains with Repeating Structure

In this section, we introduce some associated measures and results related to Markov chains with infinite state block-structured transition matrices. Results presented here are for discrete time Markov chains. However, corresponding results for continuous time Markov chains can be obtained in parallel.

Let $\{Z_n = (X_n, Y_n);\ n = 0,1,2,...\}$, with $X_n = 0,1,2,3\cdots,;\ Y_n = 1,2,\cdots,K_{X_n} < \infty$, be the Markov chain, whose transition matrix P is expressed in block matrix form:

$$P = \begin{bmatrix} P_{0,0} & P_{0,1} & P_{0,2} & \cdots & \cdots \\ P_{1,0} & P_{1,1} & P_{1,2} & \cdots & \cdots \\ P_{2,0} & P_{2,1} & P_{2,2} & \cdots & \cdots \\ \vdots & \vdots & \vdots & \vdots & \vdots \end{bmatrix}, \tag{2.42}$$

where $P_{i,j}$ is a matrix of size $K_i \times K_j$ with both $K_i < \infty$ and $K_j < \infty$. In general, P is allowed to be substochastic. Let the state space be S and partitioned accordingly into

$$S = \bigcup_{i=0}^{\infty} L_i, \tag{2.43}$$

with

$$L_i = \{(i,1),(i,2),\ldots,(i,K_i)\}. \tag{2.44}$$

In state (i,r), i is called the level variable and r the state variable. We also use the notation

$$L_{\leq i} = \bigcup_{k=0}^{i} L_k. \tag{2.45}$$

Partitioning the transition matrix P into blocks is not only done because it is convenient for comparison with results in the literature, but also because it is necessary when the Markov chain exhibits some kind of block structure. For the above Markov chain, we define matrices $R_{i,j}$ for $i < j$ and $G_{i,j}$ for $i > j$ as follows. $R_{i,j}$ is a matrix of size $K_i \times K_j$ whose $(r,\ s)$th entry is the expected number of visits to state $(j,\ s)$ before hitting any state in $L_{\leq(j-1)}$, given that the process starts in state $(i,\ r)$. $G_{i,j}$ is a matrix of size $K_i \times K_j$ whose (r,s)th entry is the probability of hitting state $(j,\ s)$ when the process enters $L_{\leq(i-1)}$ for the first time, given that the process starts in state $(i,\ r)$. We call matrices $R_{i,j}$ and $G_{i,j}$ respectively, the matrices of expected number of visits to higher levels before returning to lower levels and the matrices of the first passage probabilities to lower levels.

The significance of $R_{i,j}$ and $G_{i,j}$ in studying Markov chains, especially for ergodic Markov chains, has been pointed out in many research papers, including Zhao, Li and Alfa [107], Grassmann and Heyman [47], Grassmann and Heyman [48], and Heyman [55]). In these papers, under the ergodic condition, stationary distribution vectors, the mean first passage times and the fundamental matrix are discussed in terms of these matrices. For some recent results, see Zhao, Li and Alfa [108], Zhao, Li and Braun [107].

For special type of Markov chains in block form that have the property of repeating rows (or columns), which means that the transition probability matrix partitioned as in [108] has the following form:

$$P = \begin{bmatrix} P_{0,0} & P_{0,1} & P_{0,2} & P_{0,3} & \cdots & \cdots \\ P_{1,0} & A_0 & A_1 & A_2 & \cdots & \cdots \\ P_{2,0} & A_{-1} & A_0 & A_1 & \cdots & \cdots \\ P_{3,0} & A_{-2} & A_{-1} & A_0 & \cdots & \cdots \\ \vdots & \vdots & \vdots & \vdots & \vdots & \vdots \end{bmatrix} \quad (2.46)$$

where $P_{0,0}$ is a matrix of size $K_0 \times K_0$ and all A_v for $v = 0, \pm 1, \pm 2, \ldots$, are matrices of size $m \times m$ with $m < \infty$. The sizes of other matrices are determined accordingly. $R_{n-i} = R_{i,n}$ and $G_{n-i} = G_{n,i}$ for $i \geq 0$. We only give the following results. The details, i.e. theorems and proofs can be found in Zhao, Li and Alfa [108], Zhao, Li and Braun [107]. We define $sp(A)$ as the spectral radius of a square matrix A, i.e. its largest absolute eigenvalue.

1. Let P be stochastic.
   - Let $R = \sum_{n=1}^{\infty} R_n$. The Markov chain P in (2.46) is positive recurrent if and only if $R < \infty$ and $\lim_{k \to \infty} R^k = 0$.
   - Let $G = \sum_{n=1}^{\infty} G_n$. If $\sum_{k=-\infty}^{\infty} A_k$ is stochastic, then the Markov chain P in (2.46) is recurrent if and only if G is stochastic.
2. Let P be stochastic. If $\sum_{k=0}^{\infty} kP_{0,k}e < \infty$ and if $A = \sum_{k=-\infty}^{\infty} A_k$ is stochastic and irreducible with the stationary distribution π, then
   - P is positive recurrent if and only if $\pi R < \pi$;
   - P is null recurrent if and only if $\pi R = \pi$ and $Ge = e$; and
   - P is transient if and only if $Ge < e$.
3. If $\sum_{k=1}^{\infty} kP_{0,k}e < \infty$ then P is positive recurrent if and only if $sp(R) < 1$;
   - P is transient if and only if $sp(G) < 1$; and
   - P is null recurrent if and only if $sp(R) = sp(G) = 1$.

These are some of the general results. In what follows we focus on specific cases of infinite DTMCs.

2.9.7 Matrix-Analytical Method for Infinite State Markov Chains

The matrix-analytical method (MAM) is most suited to three classes of Markov chains, viz: i) those with the GI/M/1 structure ii) those with the M/G/1 structure, and those with the QBD structure which actually embodies the combined properties of both the GI/M1 and M/G/1 types structures. Before we go into the details, let us define the state space which is partially general for our discussions.

Consider a bivariate Markov chain $\{X_n, Y_n;\ n \geq 0\}$ such that $P_{i,j;v,k} = \Pr\{X_{n+1} = i, Y_{n+1} = j | X_n = v, Y_n = k\}$; $i, v \geq 0$, $1 \leq j,\ k \leq M$ and i and v are referred to as the levels i and phase v, (> 0), respectively. Further define $P^{(m)}_{i,j;v,k} = \Pr\{X_{n+m} = i, Y_{n+m} = j | X_n = v, Y_n = k\}$ and ${}_iP^{(m)}_{i,j;v,k} = \Pr\{X_{n+m} = i, Y_{n+m} = j | X_n = v, Y_n = k\}, X_{n+w} \neq i,\ w =$

$1,2,\cdots,m$. This probability ${}_iP^{(m)}_{i,j;v,k}$ is the taboo probability that given the DTMC starts from state (i,j) it visits state (v,k) at time m without visiting level i during that period.

2.9.7.1 The GI/M/1 Type

Let the transition matrix P be given in the block partitioned form as follows:

$$P = \begin{bmatrix} B_{00} & B_{01} & & & & \\ B_{10} & B_{11} & A_0 & & & \\ B_{20} & B_{21} & A_1 & A_0 & & \\ B_{30} & B_{31} & A_2 & A_1 & A_0 & \\ \vdots & \vdots & \vdots & \vdots & \vdots & \ddots \end{bmatrix} \tag{2.47}$$

The blocks A_{i-j+1} refer to transitions from level i to j for $i \geq 1$, $j \geq 2$, $i \geq j-1$. The blocks B_{ij} refer to transitions in the boundary areas from level i to level j for $i \geq 0$, $j = 0,1$. This transition matrix is of the GI/M/1 type and very well treated in Neuts [80]. It is called the GI/M/1 type because it has the same structure as the transition matrix of the GI/M/1 queue embedded at arrival points. It is skip-free to the right. This system will be discussed at later stages. It suffices at this point just to accept the name assigned to this DTMC.

We consider only the case for which the matrix P is irreducible and positive recurrent. Neuts [80] shows that for the matrix P to be positive recurrent, we need the condition that

$$\boldsymbol{\pi} \nu > 1 \tag{2.48}$$

where $\boldsymbol{\pi} = \boldsymbol{\pi} A$, $\boldsymbol{\pi}\mathbf{1} = 1$, $A = \sum_{i=0}^{\infty} A_i$ and $\nu = \sum_{k=0}^{\infty} kA_k\mathbf{1}$.

We define $\boldsymbol{x} = [\boldsymbol{x}_0, \boldsymbol{x}_1, \boldsymbol{x}_2, \ldots]$ as the invariant probability vector associated with P, then for $\boldsymbol{x} = \boldsymbol{x}P$, we have

$$\boldsymbol{x}_0 = \sum_{i=0}^{\infty} \boldsymbol{x}_i B_{i0}, \tag{2.49}$$

$$\boldsymbol{x}_1 = \sum_{i=0}^{\infty} \boldsymbol{x}_i B_{i1}, \tag{2.50}$$

$$\boldsymbol{x}_j = \sum_{i=0}^{\infty} \boldsymbol{x}_{j+v-1} A_v, \quad j \geq 2 \tag{2.51}$$

$$\tag{2.52}$$

Let us define a matrix R whose elements $R_{i,j}$ are given as

$$R_{j,w} = \sum_{n=0}^{\infty} {}_iP^{(n)}_{i,j;i+1,w}, \; i \geq 0, 1 \leq j \leq m, 1 \leq w \leq m.$$

The element $R_{j,w}$ is the expected number of visits to state $(i+1,w)$ before retuning to level i, given that the DTMC started in state (i,j). If the matrix P satisfies (2.48) then there exists a non-negative matrix R which has all its eigenvalues in the unit disk for which the probabilities $\boldsymbol{x}_{i+1}$ can be obtained recursively as

$$\boldsymbol{x}_{i+1} = \boldsymbol{x}_i R, \text{ or } \boldsymbol{x}_{i+1} = \boldsymbol{x}_1 R^i, \quad i \geq 1. \tag{2.53}$$

This is the matrix analogue of the scalar-geometric solution of the GI/M/1 queue and that is why it is called the matrix-geometric solution. For details and rigouros proof of this results see Neuts [80]. We present a skeletal proof only in this book. Following Neuts [80] closely, the derivation of this result can be summarized as follows:

$$P^{(n)}_{i+1,j;i+1,j} = {}_iP^{(n)}_{i+1,j;i+1,j} + \sum_{v=1}^{M}\sum_{r=0}^{n} P^{(r)}_{i+1,j;i,v} \times {}_iP^{(n-r)}_{i,v;i+1,j}, \ \forall n \geq 1. \tag{2.54}$$

As shown in Neuts [80], if we sum this equation from $n=1$ to $n=N$ and divide by N and then let $N \to \infty$, we have

$$\frac{1}{N}\sum_{n=1}^{N} P^{(n)}_{i+1,j;i+1,j}|_{N\to\infty} \to x_{i+1,j}$$

$$\frac{1}{N}\sum_{n=1}^{N} {}_iP^{(n)}_{i+1,j;i+1,j}|_{N\to\infty} \to 0$$

$$\frac{1}{N}\sum_{n=1}^{N}\sum_{r=0}^{n} P^{(r)}_{i+1,j;i,v}|_{N\to\infty} \to x_{i,v}$$

and

$$\sum_{n=1}^{N} {}_iP^{(n)}_{i,v;i+1,j}|_{N\to\infty} \to R_{v,j}$$

Hence,

$$\sum_{v=1}^{M}\frac{1}{N}\sum_{n=1}^{N}\sum_{r=0}^{n} P^{(r)}_{i+1,j;i,v}|_{N\to\infty} \to \sum_{v=1}^{M} x_{i,v}R_{v,j}$$

Neuts [80] also showed that

$$(R^q)_{j,w} = R^{(q)}_{j,w} = \sum_{n=0}^{\infty} {}_iP^{(n)}_{i,j;i+q,w}, \tag{2.55}$$

and we use this result in what follows.

The matrix R is the minimum non-negative solution to the matrix polynomial equation

$$R = \sum_{k=0}^{\infty} R^k A_k. \tag{2.56}$$

The arguments leading to this are as follows. Consider the equation for $\boldsymbol{x}_j$ given as

$$\boldsymbol{x}_j = \sum_{v=0}^{\infty} \boldsymbol{x}_{j+v-1} A_v, \quad j \geq 2$$

Now if we replace $\boldsymbol{x}_{j+v}$ with $\boldsymbol{x}_{j-1} R^{v+1}$, $v > 0,\ j > 2$, we obtain

$$\boldsymbol{x}_{j-1} R = \sum_{k=0}^{\infty} \boldsymbol{x}_{j-1} R^k A_k, \quad j \geq 2 \tag{2.57}$$

After some algebraic manipulations, we obtain Equation (2.56).

The matrix R is known as the rate matrix. The elements R_{ij} of the matrix R is the expected number of visits into state $(k+1, j)$, starting from state (k, i), until the first return to state $(k, .)$, $k > 1$. The results quoted in (2.54 and 2.55) are based on taboo probabilities and a detailed presentation of the derivations can be found in Neuts [80].

The boundary probabilities $(\boldsymbol{x}_0, \boldsymbol{x}_1)$ are determined as

$$\boldsymbol{x}_0 = \sum_{i=0}^{\infty} \boldsymbol{x}_i B_{i0} = \boldsymbol{x}_0 B_{00} + \sum_{i=1}^{\infty} \boldsymbol{x}_1 R^{i-1} B_{i0} \tag{2.58}$$

and

$$\boldsymbol{x}_1 = \sum_{i=0}^{\infty} \boldsymbol{x}_i B_{i1} = \boldsymbol{x}_0 B_{01} + \sum_{i=1}^{\infty} \boldsymbol{x}_1 R^{i-1} B_{i1} \tag{2.59}$$

When rearranged in matrix form, we obtain

$$(\boldsymbol{x}_0\ \boldsymbol{x}_1) = (\boldsymbol{x}_0, \boldsymbol{x}_1)\, B[R] \tag{2.60}$$

where

$$B[R] = \begin{bmatrix} B_{00} & B_{01} \\ \sum_{n=1}^{\infty} R^{n-1} B_{n0} & \sum_{n=1}^{\infty} R^{n-1} B_{n1} \end{bmatrix} \tag{2.61}$$

and then $(\boldsymbol{x}_0, \boldsymbol{x}_1)$ is normalized by

$$\boldsymbol{x}_0 \mathbf{1} + \boldsymbol{x}_1 (I - R)^{-1} \mathbf{1} = 1 \tag{2.62}$$

The argument behind this is that

$$\boldsymbol{x}_0 \mathbf{1} + \boldsymbol{x}_1 \mathbf{1} + \boldsymbol{x}_2 \mathbf{1} + \ldots = \boldsymbol{x}_0 \mathbf{1} + \boldsymbol{x}_1 (I + R + R^2 + \ldots) \mathbf{1} = 1$$

Because of the geometric nature of Equation (2.53), this method is popularly known as the *matrix - geometric method.*

2.9.7.2 Key Measures

The key measures usually of interest in this DTMC are

- The marginal distribution of the first variable (level) is given as

$$y_k = \boldsymbol{x}_k \mathbf{1} = \boldsymbol{x}_1 R^{k-1} \mathbf{1}, \ \ k \geq 1. \tag{2.63}$$

- The first moment is given as

$$E[X] = \boldsymbol{x}_1 [I - R]^{-2} \mathbf{1}. \tag{2.64}$$

- Tail behaviour, i.e. $p_k = Pr\{X \geq k\} = \sum_{v=k}^{\infty} \boldsymbol{x}_v \mathbf{1} = \boldsymbol{x}_1 R^{k-1} [I-R]^{-1} \mathbf{1}$. Let $\eta = sp(R)$, i.e. the spectral radius of R (i.e. its maximum absolute eigenvalue), and let $\mathbf{v}$ and $\mathbf{u}$ be the corresponding left and right eigenvectors which are normalized such that $\mathbf{u}\mathbf{1} = 1$ and $\mathbf{u}\mathbf{v} = 1$. It is known that $R^k = \eta^k \mathbf{v}\mathbf{u} + o(\eta^k), \ \ k \rightarrow \infty$. Hence $\boldsymbol{x}_j = \boldsymbol{x}_1 R^{j-1} \mathbf{1} = \kappa \eta^{j-1} + o(\eta^j), \ \ j \rightarrow \infty$, where $\kappa = \boldsymbol{x}_1 \mathbf{v}$. Hence we have

$$p_k = \kappa \eta^{k-1} (1-\eta)^{-1} + o(\eta^k), \ \ k \rightarrow \infty. \tag{2.65}$$

The key ingredient to the matrix-geometric method is the matrix R and how to compute it efficiently becomes very important especially when it has a huge dimension and its spectral radius is close to 1.

2.9.7.3 Computing matrix R

There are several methods for computing the matrix R and they are all iterative. Letting $R(v)$ be the computed value of matrix R at the vth iteration, the computation is stopped once $|R(v+1) - R(v)|_{i,j} < \varepsilon$, where ε is a small convergent criterion with values of about 10^{-12}. The simplest but not necessarily most efficient methods are the linear ones, given as:

- Given that

$$R = \sum_{k=0}^{\infty} R^k A_k$$

we can write

$$R(v+1) = \sum_{k=0}^{\infty} R^k(v) A_k, \tag{2.66}$$

with $R(0) := A_0$.

- Given that

$$R = \Big(\sum_{k=0, k \neq 1}^{\infty} R^k A_k \Big)(I - A_1)^{-1}$$

we can write

$$R(v+1)=(\sum_{k=0,k\neq 1}^{\infty} R^k(v+1)A_k)(I-A_1)^{-1}, \quad (2.67)$$

with $R(0):=A_0$.

- Given that

$$R=A_0[I-\sum_{k=1}^{\infty} R^{k-1}A_k]^{-1}$$

we can write

$$R(v+1)=A_0[I-\sum_{k=1}^{\infty} R^{k-1}(v)A_k]^{-1} \quad (2.68)$$

with $R(0):=A_0$,

More efficient methods are the cyclic reduction method by Bini and Meini [22] and the method using the non-linear programming approach by Alfa, et al. [15]. These methods are more efficient than the linear approach however, they cannot take advantage of the structures of the matrices A_k, $k=0,1,2,\cdots$, when they can be exploited.

- **NON-LINEAR PROGRAMMING APPROACH FOR** R**:** By definition, R is the minimal non-negative solution to the matrix polynomial equation $R=\sum_{k=0}^{\infty} R^k A_k$. Consider the matrix $A^*(z)=\sum_{k=0}^{\infty} A_k z^k$, $|z|\leq 1$, and let $\chi^*(z)$ be the eigenvalue of $A^*(z)$. It is known from Neuts [80] that the Perron eigenvalue of R is the smallest solution to $z=\chi^*(z)$ and that $\mathbf{u}$ is its left eigenvector. Let X be any non-negative square matrix of order m and a function $f(X)=\sum_{k=0}^{\infty} X^k A_k$. For two square matrices Y and Z of the same dimensions, let $Y \circ Z$ denote their elementwise product. Now we define a function

$$H(X)=\sum_{(i,j)=1}^{m} ([f(X)]_{i,j}-X_{i,j})^2=\mathbf{1}^T((f(X)-X)\circ(f(X)-X))\mathbf{1}. \quad (2.69)$$

It was shown in Alfa et al. [15] that if the matrix P is positive recurrent then the matrix R is the unique optimal solution to the non-linear programming problem

$$minimize \quad H(X) \quad (2.70)$$

$$subject\ to \quad \mathbf{u}X=\eta\mathbf{u}, \quad (2.71)$$

$$X\geq 0. \quad (2.72)$$

Two optimization algorithms were presented by Alfa, et al. [15] for solving this problem.

- **THE INVARIANT SUBSPACE METHOD FOR** R**:** Akar and Sohraby [1] developed this method for R matrix. The method requires that $A^*(z)=\sum_{k=0}^{\infty} z^k A_k$ be rational. However, whenever we have $A_k=0$, $K<\infty$, this condition is usually met. For most practical situations $A_k=0$, $K<\infty$, hence this method is appropriate in such situations.

2.9.7.4 Some Special Structures of the Matrix R often Encountered

The matrix-geometric method is a very convenient method to work with, and also because it has a probabilistic interpretation that makes it more meaningful with certain appeal to application oriented users. Its weakness however, is that as the size of the matrix R gets very large, the computational aspects become cumbersome. In that respect, it is usually wise to search for special structures of matrix R that can be exploited. Some of the very useful special structures are as follows:

- If A_0 is of rank one, i.e. $A_0 = \omega\boldsymbol{\beta}$, then

$$R = \omega\xi = A_0[I - \sum_{v=1}^{\infty} \eta^{v-1} A_v]^{-1}, \tag{2.73}$$

  where $\eta = \xi\omega$ is the maximal eigenvalue of the matrix R.
  The argument leading to this are based on the iteration

$$R(0) := 0,$$

$$R(k+1) := A_0(I - A_1)^{-1} + \sum_{v=2}^{\infty} R^v(k) A_v (I - A_1)^{-1}, \ k \geq 0,$$

  and letting $R(k) = \omega\xi(k)$ we have

$$\xi(1) = \boldsymbol{\beta}(I - A_1)^{-1},$$

$$\xi(k) = \boldsymbol{\beta}(I - A_1)^{-1} + \sum_{v=2}^{\infty} (\xi(k-1)\omega)^{v-1} \xi(k-1) A_v (I - A_1)^{-1}, \ k \geq 0.$$

  This leads to the explicit result given above.
- For every row of A_0 that is all zeros, the corresponding rows of R are also zeros.
- When the matrices A_i, $i \geq 0$, are of the sparse block types, it is advantageous to write out the equations of R in smaller blocks in order to reduce the computations.
- Other special structures can be found in Chakravarthy and Alfa [29], Alfa and Chakravarthy [7] and Alfa, Dolhun and Chakravarthy [8].

2.9.7.5 The M/G/1 Type:

Let the transition matrix P be given in the block partitioned form as follows:

$$P = \begin{bmatrix} B_0 & B_1 & B_2 & B_3 & \cdots \\ A_0 & A_1 & A_2 & A_3 & \cdots \\ & A_0 & A_1 & A_2 & \cdots \\ & & A_0 & A_1 & \cdots \\ & & & \vdots & \vdots \end{bmatrix}. \tag{2.74}$$

The blocks A_{j-i+1} refer to transitions from level i to level j for $j \geq i-1$. The blocks B_j refer to transitions at the boundary area from level 0 to level j. It is skip-free to the left.

The transition matrix is of the M/G/1 type and very well treated in Neuts [81].

Let $A = \sum_{k=0}^{\infty} A_k$ and let $\nu = \sum_{k=1}^{\infty} kA_k\mathbf{1}$ be finite. If A is irreducible, then there exists an invariant vector $\boldsymbol{\pi}$ such that $\boldsymbol{\pi} = \boldsymbol{\pi}A$, and $\boldsymbol{\pi}\mathbf{1} = 1$. The Markov chain P is irreducible if and only if $\rho = \boldsymbol{\pi}\nu < 1$.

If the Markov chain P is irreducible, then there exists a stochastic matrix G which is the minimal non-negative solution to the matrix polynomial equation

$$G = \sum_{k=0}^{\infty} A_k G^k \tag{2.75}$$

The elements G_{ij} of this matrix refers to the probability that the system will eventually visit state $(v+1, j)$ given that it started from state (v, i), $(v > 1)$. The arguments that lead to Equation (2.75) will be presented in a skeletal form and the rigouros proof can be found in Neuts [81].

It is clear that starting from level $i+1$ the DTMC can reach level i in one step according to the matrix A_0, or in two steps according to A_1G, i.e. it stays in level $i+1$ in the first step and then goes to level i in the second step; or in three steps according to A_2G^2, i.e. it goes up to level $i+2$ in the first step, comes down two consecutive times in the second and third steps according to $G \times G = G^2$; and this continues. Hence

$$G = A_0 + A_1G + A_2G^2 + \cdots = \sum_{\nu=}^{\infty} A_\nu G^\nu.$$

It was shown in Neuts [81] in section 2.2.1 that

- If P is irreducible, G has no zero rows and it has eigenvalue η of maximum modulus.
- If P is recurrent G is stochastic.

2.9.7.6 Stationary distribution

In order to obtain the stationary distribution for the M/G/1 system we proceed as follows.

Let $\mathbf{g} = \mathbf{g}G$, where $\boldsymbol{g1} = 1$. Further define M such that

$$M = G + \sum_{\nu=1}^{\infty} A_\nu \sum_{k=0}^{\nu-1} G^k M G^{\nu-k-1} \tag{2.76}$$

and also define $\mathbf{u}$ such that

$$\mathbf{u} = (I - G + \mathbf{1g})[I - A - (\mathbf{1} - \boldsymbol{\beta})\mathbf{g}]^{-1}\mathbf{1} \tag{2.77}$$

Further define K such that

$$K = B_0 + \sum_{v=1}^{\infty} B_v G^v \tag{2.78}$$

and let $\kappa = \kappa K$, where $\kappa \mathbf{1} = 1$. The (i, j) elements of the matrix K refer to the probability that the system will ever return to state $(0, j)$ given that is started from state $(0, i)$. Note that we say "return" because we are considering the same level 0 in both cases, so we mean a return to level 0.

If we now define the vector $x = [x_0, x_1, x_2, \ldots]$ then we can obtain x_0 as

$$x_0 = \left(\kappa \widetilde{\kappa}\right)^{-1} \kappa \tag{2.79}$$

where

$$\widetilde{\kappa} = \mathbf{1} + \sum_{v=1}^{\infty} B_v \sum_{k=0}^{v-1} G^k \mathbf{u}$$

Once x_0 is obtained the remaining x_n, $n > 1$, are obtained as follows (see Ramaswami [89]):

$$x_n = \left[x_0 \overline{B}_n + (1 - \delta_{n,1}) \sum_{j=1}^{n-1} x_j \overline{A}_{n-j+1} \right] \left(I - \overline{A}_1\right), \quad n \geq 1 \tag{2.80}$$

where

$$\overline{B}_v = \sum_{i=v}^{\infty} B_i G^{i-v} \text{ and } \overline{A}_i = \sum_{i=v}^{\infty} A_i G^{i-v}, \quad v \geq 0$$

The key ingredient to using this method is the matrix G. Just like its GI/M/1 counterpart the effort involved in its computation could be enormous if its size is huge and traffic intensity is close to 1.

2.9.7.7 Computing Matrix G

There are several methods for computing the matrix G and they are all iterative. Letting $G(v)$ be the computed value of matrix G at the vth iteration, the computation is stopped once $|G(v+1) - G(v)|_{i,j} < \varepsilon$, where ε is a small convergent criterion with values of about 10^{-12}. The simplest but not necessarily most efficient methods are the linear ones, given as:

- Given that

  $$G = \sum_{k=0}^{\infty} A_k G^k$$

  we can write

  $$G(v+1) = \sum_{k=0}^{\infty} A_k G^k(v) A_k, \tag{2.81}$$

with $G(0) := \mathbf{0}$. It has been shown that starting with a $G(0) = I$ sometimes works better.

- Given that

$$G = (I - A_1)^{-1}(\sum_{k=0,k\neq 1}^{\infty} A_k G^k)$$

we can write

$$G(v+1) = (I - A_1)^{-1}(\sum_{k=0,k\neq 1}^{\infty} A_k G^k(v+1)), \tag{2.82}$$

with $G(0) := \mathbf{0}$.

- Given that

$$G = [I - \sum_{k=1}^{\infty} A_k G^{k-1}]^{-1} A_0$$

we can write

$$G(v+1) = [I - \sum_{k=1}^{\infty} A_k G^{k-1}(v)]^{-1} A_0 \tag{2.83}$$

with $G(0) := \mathbf{0}$.

More efficient methods are the cyclic reduction method by Bini and Meini [22], subspace method by Akar and Soharby [1] and the method using the non-linear programming approach by Alfa, Sengupta and Takine [16]. These methods are more efficient than the linear approach however, they cannot take advantage of the structures of the matrices A_k, $k = 0, 1, 2, \cdots$, when they can be exploited.

- **CYCLIC REDUCTION FOR** G**:** The cyclic reduction (CR) method developed by Bini and Meini [22] capitalizes on the structure of the matrix P in trying to compute G. Note that the matrix equation for G, i.e. $G = \sum_{k=0}^{\infty} A_k G^k$ can be written as $G^v = \sum_{k=0}^{\infty} A_k G^{k+v}$, $v \geq 0$, or in matrix form as

$$\begin{bmatrix} I-A_1 & -A_2 & -A_3 & \cdots \\ -A_0 & I-A_1 & -A_2 & \cdots \\ & -A_0 & I-A_1 & \cdots \\ & & -A_0 & \cdots \\ & & & \ddots \end{bmatrix} \begin{bmatrix} G \\ G^2 \\ G^3 \\ G^4 \\ \vdots \end{bmatrix} = \begin{bmatrix} A_0 \\ 0 \\ 0 \\ 0 \\ \vdots \end{bmatrix}. \tag{2.84}$$

By using the odd-even permutation idea, used by Ye and Li [102] for the folding algorithm together with an idea similar to the Logarithmic Reduction for QBDs (to be presented later), on the block matrices we can write this matrix equation as

$$\begin{bmatrix} I-U_{1,1} & -U_{1,2} \\ -U_{2,1} & I-U_{2,2} \end{bmatrix} \begin{bmatrix} V_0 \\ V_1 \end{bmatrix} = \begin{bmatrix} 0 \\ B \end{bmatrix},$$

where

$$U_{1,1} = U_{2,2} = \begin{bmatrix} I-A_1 & -A_3 & \cdots \\ & I-A_1 & \cdots \\ & & \ddots \end{bmatrix},$$

$$U_{1,2} = \begin{bmatrix} -A_0 & -A_2 & \cdots \\ & -A_0 & \cdots \\ & & \ddots \end{bmatrix},$$

$$U_{2,1} = \begin{bmatrix} -A_2 & -A_4 & \cdots \\ -A_0 & -A_2 & \cdots \\ 0 & \ddots & \ddots \end{bmatrix}, \; B = \begin{bmatrix} A_0 \\ 0 \\ \vdots \end{bmatrix},$$

$$V_0 = G^{2k}, \; V_1 = G^{2k+1}, \; k = 1, 2, \cdots.$$

By applying standard block gaussian elimination to the above equation we obtain

$$[I - U_{2,2} - U_{2,1}(I - U_{1,1})^{-1} U_{1,2}] V_1 = B. \tag{2.85}$$

By noticing that this equation (2.85) is also of the form

$$\begin{bmatrix} I - A_1^{(1)} & -A_2^{(1)} & -A_3^{(1)} & \cdots \\ -A_0^{(1)} & I - A_1^{(1)} & -A_2^{(1)} & \cdots \\ & -A_0^{(1)} & I - A_1^{(1)} & \cdots \\ & & -A_0^{(1)} & \cdots \\ & & & \ddots \end{bmatrix} \begin{bmatrix} G \\ G^3 \\ G^5 \\ G^7 \\ \vdots \end{bmatrix} = \begin{bmatrix} A_0 \\ 0 \\ 0 \\ 0 \\ \vdots \end{bmatrix},$$

Bini and Meini [22] showed that after the nth repeated application of this operation we have

$$\begin{bmatrix} I - U_{1,1}^{(n)} & -U_{1,2}^{(n)} \\ -U_{2,1}^{(n)} & I - U_{2,2}^{(n)} \end{bmatrix} \begin{bmatrix} V_0^{(n)} \\ V_1^{(n)} \end{bmatrix} = \begin{bmatrix} 0 \\ B \end{bmatrix},$$

where

$$U_{1,1}^{(n)} = U_{2,2}^{(n)} = \begin{bmatrix} I - A_1^{(n)} & -A_3^{(n)} & \cdots \\ & I - A_1^{(n)} & \cdots \\ & & \ddots \end{bmatrix},$$

$$U_{1,2}^{(n)} = \begin{bmatrix} -A_0^{(n)} & -A_2^{(n)} & \cdots \\ & -A_0^{(n)} & \cdots \\ & & \ddots \end{bmatrix},$$

$$U_{2,1}^{(n)} = \begin{bmatrix} -A_2^{(n)} & -A_4^{(n)} & \cdots \\ -A_0^{(n)} & -A_2^{(n)} & \cdots \\ 0 & \ddots & \ddots \end{bmatrix}, \; B = \begin{bmatrix} A_0 \\ 0 \\ \vdots \end{bmatrix},$$

$$V_0 = G^{2k.2^n}, \; V_1 = G^{2k.2^n+1}, \; k = 1, 2, \cdots, \; n \geq 0.$$

In the end they show that the matrix equation is also of the form

$$[I - U_{2,2}^{(n)} - U_{2,1}^{(n)}(I - U_{1,1}^{(n)})^{-1}U_{1,2}^{(n)}]V_1^{(n)} = B,$$

which results in

$$G = (I - A_1^{(n)})^{-1}(A_0 + \sum_{k=1}^{\infty} A_k^{(n)} G^{k.2^n+1}). \tag{2.86}$$

They further showed that as $n \rightarrow \infty$ the second term in the brackets tends to zero, and $(I - A_1^{(n)})^{-1}$ exists, hence

$$G = (I - A_1^{(\infty)})^{-1}A_0. \tag{2.87}$$

Hence G can be obtained from this expression. Of course the operation can only be applied in a finite number of times. So it is a matter of selecting how many operations to carry out given the tolerance desired.

- **THE INVARIANT SUBSPACE METHOD FOR** G**:** Akar and Sohraby [1] developed this method for G matrix. The method requires that $A^*(z) = \sum_{k=0}^{\infty} z^k A_k$ be rational. However, whenever we have $A_k = 0,\ K < \infty$, this condition is usually met. For most practical situations $A_k = 0,\ K < \infty$, hence this method is appropriate in such situations.
- **NON-LINEAR PROGRAMMING APPROACH FOR** G**:** The non-linear programming for the matrix G is similar in concept to that used for the matrix R, with some minor differences. Consider an $m \times m$ matrix of non-negative real values such that $g(X) = \sum_{k=0}^{\infty} A_k X^k$, where $g : \Re^m \rightarrow \Re^m$. Further define a function $\phi(X)$ such that $\phi : \Re^m \rightarrow \Re^1$ and it is order preserving, i.e. $X < Y$ implies $\phi(X) < \phi(Y)$ where both $(X,Y) \in \Re^m$. A good example is $\phi(X) = \mathbf{1}^T X \mathbf{1}$. It was shown by Alfa, et al. [16] that the matrix G is the unique and optimal solution to the non-linear programming problem

$$minimize \quad \phi(X) \tag{2.88}$$

$$subject\ to \quad g(X) - X \leq 0, \tag{2.89}$$

$$X \geq 0. \tag{2.90}$$

Several algorithms were proposed by Alfa, et al. [16] for solving this problem.

2.9.7.8 Some Special Structures of the Matrix G often Encountered

The matrix-geometric method is a very convenient method to work with, and also because it has a probabilistic interpretation that makes it more meaningful with certain appeal to application oriented users. Its weakness is that as the size of the matrix G gets very large, the computational aspects become cumbersome. In that respect, it is usually wise to search for special structures of matrix G that can be exploited. Some of the very useful special structures are as follows:

- If A_0 is of rank one, i.e. $A_0 = \mathbf{v}\boldsymbol{\alpha}$, with $\boldsymbol{\alpha}\mathbf{1} = 1$, then

$$G = \mathbf{1}\boldsymbol{\alpha}, \tag{2.91}$$

if the Markov chain with transition matrix P is positive recurrent. This is obtained through the following arguments. Note that

$$G(1) := A_0,$$

$$G(k+1) := \sum_{\ell=0}^{\infty} A_\ell G(k)^\ell, \; k \geq 1.$$

If we write $G(k) = \gamma(k)\boldsymbol{\alpha}, \;\; k \geq 1$, where

$$\gamma(1) = \mathbf{v},$$

$$\gamma(k+1) = \sum_{\ell=0}^{\infty} A_\ell(\boldsymbol{\alpha}\gamma(k-1))^{\ell-1}\gamma(k-1).$$

Hence we have $G = \gamma\boldsymbol{\alpha}$, and since the Markov chain is positive recurrent we have $G\mathbf{1} = \mathbf{1}$, implying that $\gamma = \mathbf{1}$ and hence

$$G = \mathbf{1}\boldsymbol{\alpha}.$$

- For every column of A_0 that is all zeros, the corresponding columns of G are also all zeros.
- When the matrices A_i, $i \geq 0$, are of the sparse block types, it is advantageous to write out the equations of G in smaller blocks in order to reduce the computations.
- Other special structures can be found in Chakravarthy and Alfa [29], Alfa and Chakravarthy [7] and Alfa, Dolhun and Chakravarthy [8].

2.9.7.9 QBD

The QBD (Quasi-Birth-and-Death) DTMC is the most commonly encountered DTMC in discrete time queues. It embodies the properties of both the GI/M/1 and M/G/1 types of DTMCs. This infinite DTMC will be dealt with in more details because of its importance in queueing theory.

Let the transition matrix P be given in the block partitioned form as follows:

$$P = \begin{bmatrix} B & C & & & \\ E & A_1 & A_0 & & \\ & A_2 & A_1 & A_0 & \\ & & A_2 & A_1 & A_0 \\ & & & \ddots & \ddots & \ddots \end{bmatrix}. \tag{2.92}$$

The QBD only goes a maximum of one level up or down. It is skip-free to the left and to the right. These are very useful properties which are fully exploited in the theory of QBDs. For detailed treatment of this see Latouche and Ramaswami [67].

We assume that the matrices A_k, $k = 0,1,2$ are of dimension $n \times n$ and matrix B is of dimension $m \times m$, hence C and E are of dimensions $m \times n$ and $n \times m$, respectively. The key matrices that form the ingredients for analyzing a QBD are the R and G matrices, which are given as

$$R = A_0 + RA_1 + R^2 A_2 \tag{2.93}$$

and

$$G = A_2 + A_1 G + A_0 G^2. \tag{2.94}$$

R and G are the minimal non-negative solutions to the first and second equation, respectively. Another matrix that is of importance is the matrix U, which records the probability of visiting level i before level $i-1$, given the DTMC started from level i. The matrix U is given as

$$U = A_1 + A_0(I-U)^{-1}A_2. \tag{2.95}$$

There are known relationships between the three matrices R, G and U, and are given as follows

$$R = A_0(I-U)^{-1}, \tag{2.96}$$

$$G = (I-U)^{-1}A_2, \tag{2.97}$$

$$U = A_1 + A_0 G, \tag{2.98}$$

$$U = A_1 + RA_2. \tag{2.99}$$

Using these known relationships we have

$$R = A_0(I - A_1 - A_0 G)^{-1}, \tag{2.100}$$

$$G = (I - A_1 - RA_2)^{-1}A_2. \tag{2.101}$$

These last two equations are very useful for obtaining R from G and vice-versa, especially when it is easier to compute one of them when our interest is in the other.

In what follows we present a skeletal development of the relationship between the vector $\boldsymbol{x}_{i+1}$ and $\boldsymbol{x}_i$ through the matrix R using a different approach from the one used for the GI/M/1 system. The method is based on censoring used differently by Latouche and Ramaswami [67] and Grassmann and Heyman [48]. Consider the matrix P above. It can be partitioned as follows

$$P = \begin{bmatrix} T_L & T_R \\ B_L & B_R \end{bmatrix}. \tag{2.102}$$

Suppose we select n from $0,1,2,\cdots$ as the point of partitioning. Then the vector $\boldsymbol{x}$ is also partitioned into $\boldsymbol{x} = [\boldsymbol{x}_T \ \boldsymbol{x}_B]$, with

$$\boldsymbol{x}_T = [\boldsymbol{x}_0, \boldsymbol{x}_1, \cdots, \boldsymbol{x}_n], \ \boldsymbol{x}_B = [\boldsymbol{x}_{n+1}, \boldsymbol{x}_{n+2}, \cdots].$$

The result is that we have

$$T_R = \begin{bmatrix} 0 & 0 & 0 & \cdots \\ \vdots & \vdots & \vdots & \vdots \\ 0 & 0 & 0 & \cdots \\ A_0 & 0 & 0 & \cdots \end{bmatrix}, \tag{2.103}$$

and

$$B_R = \begin{bmatrix} A_1 & A_0 & & \\ A_2 & A_1 & A_0 & \\ & A_2 & A_1 & A_0 \\ & & \ddots & \ddots & \ddots \end{bmatrix}. \tag{2.104}$$

Keeping in mind that $\boldsymbol{x} = \boldsymbol{x}P$, we can write

$$\boldsymbol{x}_T = \boldsymbol{x}_T T_L + \boldsymbol{x}_B B_L \tag{2.105}$$

$$\boldsymbol{x}_B = \boldsymbol{x}_T T_R + \boldsymbol{x}_B B_R. \tag{2.106}$$

Equation (2.106) can be written as

$$\boldsymbol{x}_B = \boldsymbol{x}_T T_R (I - B_R)^{-1}, \tag{2.107}$$

which can be further written as

$$[\boldsymbol{x}_{n+1}, \boldsymbol{x}_{n+2}, \cdots] = [\boldsymbol{x}_0, \boldsymbol{x}_1, \cdots, \boldsymbol{x}_n] T_R (I - B_R)^{-1}.$$

By the structure of T_R we know that the structure of $T_R(I - B_R)^{-1}$ is as follows

$$T_R(I - B_R)^{-1} = \begin{bmatrix} 0 & 0 & 0 & \cdots \\ \vdots & \vdots & \vdots & \vdots \\ 0 & 0 & 0 & \cdots \\ A_0 V_{11} & A_0 V_{12} & A_0 V_{13} & \cdots \end{bmatrix}, \tag{2.108}$$

where V_{ij} are the block elements of $(I - B_R)^{-1}$. It is clear that

$$\boldsymbol{x}_{n+1} = \boldsymbol{x}_n A_0 V_{11} = \boldsymbol{x}_n R, \tag{2.109}$$

where R is by definition equivalent to $A_0 V_{11}$. This is another way to see the matrix geometric result for the QBD. See Latouche and Ramaswami [67] for details.

Similar to the GI/M/1 type, if the QBD DTMC is positive recurrent then the matrix R has a spectral radius less than 1 and also the matrix G is stochastic just like the case for the M/G/1 type. Once we know R we can use the result $\boldsymbol{x}_{i+1} = \boldsymbol{x}_i R, \ i \geq 1$. However, we still need to compute R and determine the boundary behaviour of the matrix P. First we discuss the boundary behaviour.

The censored stochastic transition matrix representing the boundary is $B[R]$ and can be written as

$$B[R] = \begin{bmatrix} B & C \\ E & A_1 + RA_2 \end{bmatrix}.$$

Using this matrix we obtain

$$\boldsymbol{x}_1 = \boldsymbol{x}_1[E(I-B)^{-1}C + A_1 + RA_2],$$

which after getting solved for $\boldsymbol{x}_1$ we can then solve for $\boldsymbol{x}_0$ as $\boldsymbol{x}_0 = \boldsymbol{x}_1 E(I-B)^{-1}$, and then normalized through

$$\boldsymbol{x}_0\mathbf{1} + \boldsymbol{x}_1(I-R)^{-1}\mathbf{1} = 1.$$

2.9.7.10 Computing the matrices R and G

Solving for R and G can be carried out using the techniques for the GI/M/1 type and M/G/1 type respectively. However, very efficient methods have been developed for the QBD, such as

- **THE LOGARITHMIC REDUCTION (LR) METHOD:** This is a method developed by Latouche and Ramaswami [65] for QBDs and is quadratically convergent. The Cyclic Reduction method discussed earlier is similar in principle to the Logarithmic Reduction method. The LR method is vastly used for analyzing QBDs. Readers are referred to the original paper [65] and the book [67], both by Latouche and Ramaswami, for detail coverage of the method. In this section we present the results and give an outline of the algorithm. Essentially the algorithm works on computing the matrix G by kind of restricting the DTMC to only time epochs when it changes main levels, i.e. it goes up or down one, hence the transitions without a level change are censored out. Once the matrix G is obtained, the matrix R can be easily calculated from the relationship given earlier between R and G.
  Consider the transition matrix P and write the G matrix equation given as

  $$G = A_2 + A_1 G + A_0 G^2.$$

  This can be written as $(I-A_1)G = A_2 + A_0 G^2$ or better still as

  $$G = (I-A_1)^{-1}A_2 + (I-A_1)^{-1}A_0 G^2 = L + HG^2, \tag{2.110}$$

  with

  $$L = (I-A_1)^{-1}A_2, \text{ and } H = (I-A_1)^{-1}A_0. \tag{2.111}$$

  If we restrict the process to only up and down movements while not considering when there is no movement, then the matrix L captures its down movement while the matrix H captures its up movements. Note that G and G^2 record first passage probabilities across one and two levels, respectively. So if we repeat this process we can study first passage probabilities across 4, 8, $\cdots$ levels. So let us proceed and define

$$A_k^{(0)} = A_k,\ k = 0,1,2$$

$$H^{(0)} = (I - A_1^{(0)})^{-1} A_0^{(0)},\ L^{(0)} = (I - A_1^{(0)})^{-1} A_2^{(0)},$$

$$U^{(k)} = H^{(k)} L^{(k)} + L^{(k)} H^{(k)},\ \ k \geq 0,$$

$$H^{(k+1)} = (I - U^{(k)})^{-1} (H^{(k)})^2,\ L^{(k+1)} = (I - U^{(k)})^{-1} (L^{(k)})^2,\ k \geq 0.$$

After aplying this process repeatedly Latouche and Ramaswami [65] showed that

$$G = \sum_{k=0}^{\infty} \left(\prod_{i=0}^{k-1} H^{(i)} \right) L^{(k)}. \tag{2.112}$$

Since we can not compute the sum series infinitely, the sum is truncated at some point depending on the predefined tolerance.

- **THE CYCLIC REDUCTION METHOD:** The cyclic reduction method developed by Bini and Meini [22] for the M/G/1 is another appropriate algorithm for the QBD in the same sense as the LR. We compute the matrix G and use that to compute the matrix R. The algorithm will not be repeated here since it has been presented under the M/G/1 techniques.
- **THE INVARIANT SUBSPACE METHOD:** Akar and Sohraby [1] developed this method for G and R matrices, and since for the QBDs $A(z)$ are rational this method is appropriate for such analysis.

2.9.7.11 Some Special Structures of the Matrix R and the matrix G

If A_2 or A_0 is of rank one, then the matrix R can be obtained explicitly as the inverse of another matrix, and G matrix can also be obtained as product of a column vector and a row vector, as follows.

- if $A_2 = \mathbf{v}.\boldsymbol{\alpha}$, then $R = A_0[I - A_1 - A_0 G]^{-1}$, where $G = \mathbf{1}.\boldsymbol{\alpha}$,
- if $A_0 = \omega.\boldsymbol{\beta}$, then $R = A_0[I - A_1 - \eta A_2]^{-1}$, where $\xi = \boldsymbol{\beta}[I - A_1 - \eta A_2]^{-1}$, $\xi\omega = \eta$, $\xi A_2 \mathbf{1} = 1$ and the matrix R satisfies $R^i = \eta^{i-1} R$, $i \geq 1$.
- All the special structures identified for the M/G/1 and GI/M/1 types are also observed for the QBDs.

2.9.8 *Other special QBDs of interest*

2.9.8.1 Level-dependent QBDs:

A level dependent QBD has a transition matrix P given in the block partitioned form as follows:

$$P = \begin{bmatrix} A_{0,0} & A_{0,1} & & & \\ A_{1,0} & A_{1,1} & A_{1,2} & & \\ & A_{2,1} & A_{2,2} & A_{2,3} & \\ & & A_{3,2} & A_{3,3} & A_{3,4} \\ & & & \ddots & \ddots & \ddots \end{bmatrix}, \tag{2.113}$$

We assume that the matrices $A_{k,k},\ k = 0,1,2$ are of dimension $m_k \times m_k$ and the dimensions of the matrices $A_{k,k+1}$ are $m_k \times m_{k+1}$ and of $A_{k+1,k}$ are $m_{k+1} \times m_k$. Because the transition matrix is level dependent, the matrices R and G are now level dependent. We have R_k which records the rate of visiting level k before coming back to level $k-1$, given it started from level $k-1$, and G_k records the probability of first visit from level k to level $k-1$. The matrices are obtained as

$$R_k = A_{k-1,k} + R_k A_{k,k} + R_k R_{k+1} A_{k+1,k}, \tag{2.114}$$

and

$$G_k = A_{k,k-1} + A_{k,k} G_k + A_{k,k+1} G_{k+1} G_k. \tag{2.115}$$

For this level dependent QBD we have matrix-product solution given as

$$\boldsymbol{x}_{k+1} = \boldsymbol{x}_k R_k,\ \ k \geq 0. \tag{2.116}$$

The condition for this QBD to be positive recurrent is given in Latouche and Ramaswami [67] as follows:

Condition: That

$$\boldsymbol{x}_0 = \boldsymbol{x}_0 (A_{0,0} + A_{0,1} G_1), \tag{2.117}$$

with

$$\boldsymbol{x}_0 \sum_{n=0}^{\infty} \prod_{k=1}^{n} R_k \mathbf{1} = 1. \tag{2.118}$$

The level dependent QBD is usually difficult to implement in an algorithm form in general. However, special cases can be dealt with using specific features of the DTMC. For more details on level dependent QBDs and algorithms see Latouche and Ramaswami [67]. Some of the special cases will be presented later in the queueing section of this book, especially when we present some classes of vacation models.

2.9.8.2 Tree structured QBDS

Consider a DTMC $\{(X_n, Y_n), n \geq 0\}$ in which X_n assumes values of the nodes of a d-ary tree, and Y_n as the auxilliary variables such as phases. As an example a 2-ary has for level 1 the set $\{1,2\}$, for level 2 the set $\{11,12,21,22\}$ for level 3 the set $\{111,112,121,122,211,212,221,222\}$ and so on. If we now allow this DTMC to go up and down not more than one level, we then have a tree structured QBD. Details of such DTMC can be found in Yeung and Alfa [103]. We can also have the GI/M/1 type and M/G/1 type of tree structured DTMC. For further reference see Yeung and Sengupta [104] and Takine, Sengupta and Yeung [95], respectively.

A d-ary tree is a tree which has d children at each node, and the root of the tree is labelled $\{0\}$. The rest of the nodes are labelled as strings. Each node, which represents a level say i in a DTMC, has a string of length i with each element of the string consisting of a value in the set $\{1,2,\cdots,d\}$. We use $+$ sign in this section to represent concatenation on the right. Consider the case of $d=2$ and for example if we let $J=1121$ be a node and $k=2$ then $J+k=11212$. We adopt the convention of using upper case letters for strings and lower case letters for integers. Because we are dealing with a QBD case we have that the DTMC, when in node $J+k$ could only move in one step to node J or node $J+k+s$ or remain in node $J+k$, where $k,s=1,2,\cdots,d$. The last come first served single server queues have been analysed using this tree structured QBD.

For this type of QBD most of the results for the standard QBD still follow with only minor modifications to the notations. Consider the chain in a node $J+k$ and phase i at time n, i.e. the DTMC is in state $(J+k,i)$, where i is an auxilliay variable $i=1,2,\cdots,m$. At time $n+1$ the DTMC could be at state:

- (J,j) with probability $d_k^{i,j}$, $k=1,2,\cdots,d$
- $(J+s,j)$ with probability $a_{k,s}^{i,j}$, $k,s=1,2,\cdots,d$
- $(J+ks,j)$ with probability $u_s^{i,j}$, $k,s=1,2,\cdots,d$

Let G_k be an $m\times m$ matrix which records the probability that the DTMC goes from node $J+k$ to node J for the first time. This is really equivalent to the G matrix in the standard QBD and to the G_k matrix in the level dependent QBD. If we define $m\times m$ matrices D_k, $A_{k,s}$, and U_s as the matrices with (i,j) elements $d_k^{i,j}$, $a_{k,s}^{i,j}$ and $u_s^{i,j}$, then we have

$$G_k = D_k + \sum_{s=1}^{d} A_{k,s}G_s + \sum_{s=1}^{d} U_sG_sG_k, \quad k=1,2,\cdots,d. \tag{2.119}$$

Note that

$$(D_k + \sum_{s=1}^{d} A_{k,s} + \sum_{s=1}^{d} U_s)\mathbf{1} = \mathbf{1}. \tag{2.120}$$

Similarly, if we define R_k as the matrix that records the expected number of visits to node $(J+k)$ before visiting node J, given that it started from node J, then we have

$$R_k = U_k + \sum_{s=1}^{d} R_sA_{s,k} + \sum_{s=1}^{d} R_kR_sD_s, \quad k=1,2,\cdots,d. \tag{2.121}$$

Because the summations operations in the equations for both the G_k and R_k matrices are finite, we can simply apply the linear approach for the standard R and G matrices to compute these G_k and R_k matrices.

Provided the system is stable we can apply the linear algorithm as follows:

- Set

$$G_k(0) := D_k, \;\; R_k(0) := U_k, \;\; k = 1,2,\cdots,d$$

- Compute

$$G_k(n+1) := D_k + \sum_{s=1}^{d} A_{k,s} G_s(n) + \sum_{s=1}^{d} U_s G_s(n) G_k(n), \; k = 1,2,\cdots,d,$$

$$R_k(n+1) := U_k + \sum_{s=1}^{d} R_s(n) A_{s,k} + \sum_{s=1}^{d} R_k(n) R_s(n) D_s, \; k = 1,2,\cdots,d.$$

- Stop if

$$(G_k(n+1) - G_k(n))_{i,j} < \varepsilon, \forall i,j,k,$$

and

$$(R_k(n+1) - R_k(n))_{i,j} < \varepsilon, \forall i,j,k,$$

where ε is a preselected tolerance value.

2.9.9 Re-blocking of transition matrices

For most practrical situations encountered we find that the resulting transition matrices do not have exactly any of the three structures presented. However, in most cases the structures are close to one of the three and by re-blocking the transition matrix we can achieve one of them

2.9.9.1 The non-skip-free M/G/1 type

As an example, we can end up with a non-skip-free M/G/1 type DTMC with a transition matrix of the following form

$$P = \begin{bmatrix} C_{0,0} & C_{0,1} & C_{0,2} & C_{0,3} & C_{0,4} & \cdots \\ C_{1,0} & C_{1,1} & C_{1,2} & C_{1,3} & C_{1,4} & \cdots \\ \vdots & \vdots & \vdots & \vdots & \vdots & \cdots \\ C_{k-1,0} & C_{k-1,1} & C_{k-1,2} & C_{k-1,3} & C_{k-1,4} & \cdots \\ A_0 & A_1 & A_2 & A_3 & A_4 & \cdots \\ & A_0 & A_1 & A_2 & A_3 & \cdots \\ & & A_0 & A_1 & A_2 & \cdots \\ & & & \ddots & \ddots & \ddots \end{bmatrix}. \quad (2.122)$$

This can be re-blocked further into $k \times k$ superblocks so that we have

$$P = \begin{bmatrix} C_0 & C_1 & C_2 & C_3 & \cdots & \\ H_0 & H_1 & H_2 & H_3 & \cdots & \\ & H_0 & H_1 & H_2 & \cdots & \\ & & H_0 & H_1 & \cdots & \\ & & & \ddots & \ddots & \ddots \end{bmatrix}, \tag{2.123}$$

where

$$C_j = \begin{bmatrix} C_{0,jk} & \cdots & C_{0,(j+1)k-1} \\ \vdots & \cdots & \vdots \\ C_{k-1,jk} & \cdots & C_{k-1,(j+1)k-1} \end{bmatrix}, \; j = 0,1,2,\cdots$$

$$H_j = \begin{bmatrix} A_{jk} & \cdots & A_{(j+1)k-1} \\ \vdots & \cdots & \vdots \\ A_{(j-1)k} & \cdots & A_{jk} \end{bmatrix}, \; j = 1,2,\cdots,$$

and

$$H_0 = \begin{bmatrix} A_0 & A_1 & \cdots & A_{k-1} \\ & A_0 & \cdots & A_{k-2} \\ & & \ddots & \vdots \\ & & & A_0 \end{bmatrix}.$$

This non-skip-free M/G/1 type DTMC has now been converted to an M/G/1 type which is skip-free to the left. Standard M/G/1 results and techniques can now be applied to analyze this DTMC. We point out that there is a class of algorithms specifically designed for this class of problems by Gail, Hantler and Taylor [43].

Suppose we have a situation where $C_j = 0$ and $H_{j+w} = 0$, $\forall j > M$, $0 \le w < \infty$, then we can further reblock this transition matrix to the QBD type as follows. We display the case of $w = 0$ only. Let

$$B = C_0, \; C = [C_1, \; C_2, \; \cdots, \; C_M], \; E = \begin{bmatrix} H_0 \\ 0 \\ \vdots \\ 0 \end{bmatrix},$$

$$U = \begin{bmatrix} 0 & 0 & 0 & \cdots & 0 \\ H_M & 0 & 0 & \cdots & 0 \\ H_{M-1} & H_M & \ddots & \ddots & 0 \\ \vdots & \vdots & \ddots & \vdots & \cdots \\ H_1 & H_2 & \cdots & H_M & 0 \end{bmatrix}, \quad D = (\mathbf{e}_1 \otimes \mathbf{e}_M^T) \otimes H_0,$$

$$A = \begin{bmatrix} H_1 & H_2 & H_3 & \cdots & H_M \\ H_0 & H_1 & H_2 & \cdots & H_{M-1} \\ & H_0 & H_1 & \cdots & H_{M-2} \\ & & \ddots & \ddots & \vdots \\ & & & H_0 & H_1 \end{bmatrix},$$

where $\mathbf{e}_j$ is a column vector of zeros with one in location j and $\mathbf{e}_j^T$ is its transpose, then we can write the transition matrix as

$$P = \begin{bmatrix} B & C & & & \\ E & A & U & & \\ & D & A & U & \\ & & D & A & U \end{bmatrix},$$

which is of the QBD type. In this case we can apply the results of QBD to analyze the system.

2.9.9.2 The non-skip-free GI/M/1 type

Similar to the M/G/1 type DTMC we can have a GI/M/1 non-skip-free type of DTMC with transition matrix of the form

$$P = \begin{bmatrix} B_{0,0} & B_{0,1} & \cdots & B_{0,k-1} & A_0 & & & \\ B_{1,0} & B_{1,1} & \cdots & B_{1,k-1} & A_1 & A_0 & & \\ B_{2,0} & B_{2,1} & \cdots & B_{2,k-1} & A_2 & A_1 & A_0 & \\ \vdots & \vdots & \cdots & \vdots & \vdots & \vdots & \vdots & \ddots \end{bmatrix}. \tag{2.124}$$

This can also be re-blocked further into $k \times k$ superblocks to obtain

$$P = \begin{bmatrix} B_0 & F_0 & & & \\ B_1 & F_1 & F_0 & & \\ B_2 & F_2 & F_1 & F_0 & \\ \vdots & \vdots & \vdots & \vdots & \ddots \end{bmatrix}, \tag{2.125}$$

where

$$B_j = \begin{bmatrix} B_{jk,0} & \cdots & B_{jk,k-1} \\ \vdots & \cdots & \vdots \\ B_{(j+1)k-1,0} & \cdots & B_{(j+1)k-1,k-1} \end{bmatrix}, \; j = 0, 1, 2, \cdots,$$

$$H_j = \begin{bmatrix} A_{jk} & \cdots & A_{(j-1)k} \\ \vdots & \cdots & \vdots \\ A_{(j+1)k-1} & \cdots & A_{jk} \end{bmatrix}, \; j = 1, 2, \cdots,$$

$$H_0 = \begin{bmatrix} A_0 & & & \\ A_1 & A_0 & & \\ \vdots & \vdots & \ddots & \\ A_{k-1} & A_{k-2} & \cdots & A_0 \end{bmatrix}.$$

This non-skip-free DTMC has now been converted to a GI/M/1 type which is skip-free to the right. Standard GI/M/1 results can now be applied to analyze this DTMC. Once again we point out that there is a class of algorithms specifically designed for this class of problems by Gail, Hantler and Taylor [43].

Suppose we have a situation where $B_j = 0$ and $F_{j+v} = 0,\ \forall j > N,\ 0 \le v < \infty$, then we can further reblock this transition matrix to the QBD type as follows. We display the case of $v = 0$ only. Let

$$B = B_0,\ C = [F_0,\ 0,\ 0,\ \cdots,\ 0],\ E = \begin{bmatrix} B_1 \\ B_2 \\ \vdots \\ B_N \end{bmatrix},$$

$$A = \begin{bmatrix} F_1 & F_0 & & & \\ F_2 & F_1 & F_0 & & \\ F_3 & F_2 & F_1 & F_0 & \\ \vdots & \vdots & \vdots & \ddots & \ddots \\ F_N & F_{N-1} & F_{N-2} & \cdots & F_1 \end{bmatrix},$$

$$D = \begin{bmatrix} 0 & F_N & F_{N-1} & \cdots & F_2 \\ & & F_N & \cdots & F_3 \\ & & & \ddots & \vdots \\ & & & & F_N \\ & & & & 0 \end{bmatrix},\quad U = (\mathbf{e}_1 \otimes \mathbf{e}_1^T) \otimes F_1,$$

then we can write the transition matrix as

$$P = \begin{bmatrix} B & C & & & \\ E & A & U & & \\ & D & A & U & \\ & & D & A & U \end{bmatrix},$$

which is of the QBD type. In this case we can apply the results of QBD to analyze the system.

In general a banded transition matrix can be reduced to a QBD in structure.

2.9.9.3 Time-inhomogeneous Discrete Time Markov Chains

Up till now we have focussed mainly on Markov chains that are independent of time, i.e. where $Pr\{X_{n+1} = j|X_n = i\} = p_{i,j}$. Here we deal with cases where $Pr\{X_{n+1} = j|X_n = i\} = p^n_{i,j}$, i.e. the transition probability depends on the time of the transition. We present the case of time-inhomogeneous QBDs here. The case of the GI/M/1 and M/G/1 types have been discussed in details by Alfa and Margolius [2].

For the time-inhomogeneous QBDs we let the bivariate process be $\{X_n, J_n\}, n \geq 0$ with $X_n \geq 0,\ 1 \leq J_n \leq M < \infty$. We can write the associated transition matrix as

$$P^{(n)} = \begin{bmatrix} A^{(n)}_{0,0} & A^{(n)}_{0,1} & & & \\ A^{(n)}_{1,0} & A^{(n)}_{1,1} & A^{(n)}_{1,2} & & \\ & A^{(n)}_{2,1} & A^{(n)}_{2,2} & A^{(n)}_{2,3} & \\ & & \ddots & \ddots & \ddots \end{bmatrix}, \tag{2.126}$$

where the matrices $A^{(n)}_{i,j}$ records the transition from time n to time $n+1$ with X_n changing from i to j. Hence $(A^{(n)}_{i,j})_{\ell,k} = Pr\{X_{n+1} = j, J_{n+1} = k|X_n = i, J_n = \ell\}$. If we now define $(A^{(n,m)}_{i,j})_{\ell,k} = Pr\{X_{m+1} = j, J_{m+1} = k|X_n = i, J_n = \ell\}, m \geq n$. Note that we have $A^{(n,n)} = A^{(n)}$. Let the corresponding transition matrix be $P^{(n,m)}$. For simplicity we only display the case of $m = n+1$. We can write

$$P^{(n,n+1)} = \begin{bmatrix} A^{(n,n+1)}_{0,0} & A^{(n,n+1)}_{0,1} & A^{(n,n+1)}_{0,2} & & & & \\ A^{(n,n+1)}_{1,0} & A^{(n,n+1)}_{1,1} & A^{(n,n+1)}_{1,2} & A^{(n,n+1)}_{1,3} & & & \\ A^{(n,n+1)}_{2,0} & A^{(n,n+1)}_{2,1} & A^{(n,n+1)}_{2,2} & A^{(n,n+1)}_{2,3} & A^{(n,n+1)}_{2,4} & & \\ & A^{(n,n+1)}_{3,1} & A^{(n,n+1)}_{3,2} & A^{(n,n+1)}_{3,3} & A^{(n,n+1)}_{3,4} & A^{(n,n+1)}_{3,5} & \\ & \ddots & \ddots & \ddots & \ddots & \ddots \end{bmatrix}, \tag{2.127}$$

where $A^{(n,n+1)}_{i,j} = \sum_{v=i-1}^{i+1} A^{(n)}_{i,v} A^{(n)}_{v,j}$, keeping in mind that $A^{(m)}_{\ell,k} = 0,\ \forall|\ell - k| > 1,\ m \geq 0$. Also note that $A^{(n,n+1)}_{i,j} = 0,\ \forall|i-j| > 2$. For the general case of $P^{(n,m)}$, the block matrices $A^{(n,m)}_{i,j}$ can be obtained through recursion as

$$A^{(n,m+1)}_{i,j} = \sum_{v=i-m-1}^{i+1} A^{(n,m)}_{i,v} A^{(m)}_{v,j}, \tag{2.128}$$

keeping in mind that some of the block matrices inside the summation will have zero values only and $A^{(n,m)}_{i,j} = 0,\ \forall|i-j| > m-n+1$..

If we know the state of the system at time n we can easily determine its state at time $m+1$. For example, let $\boldsymbol{x}_n = [\boldsymbol{x}^{(n)}_0,\ \boldsymbol{x}^{(n)}_1,\ \cdots]$ and $\boldsymbol{x}^{(n)}_i = [x^{(n)}_{i,1},\ x^{(n)}_{i,2},\ \cdots,\ x^{(n)}_{i,M}]$, then we have

$$\boldsymbol{x}^{(m+1)} = \boldsymbol{x}^{(n)} P^{(n,m)},\ 0 \leq n \leq m. \tag{2.129}$$

Usually of interest to us in this system is the case that we call "periodic", i.e. where we have for some integer $\tau \geq 1$ and thus $P^{(n)} = P^{(n+\tau)}$. In this case we have

$$P^{(n+K\tau,n+(K+1)\tau-1)} = P^{(n,n+\tau-1)}, \ \forall K > 0. \tag{2.130}$$

Let $\mathscr{P}^{(n)} = P^{(n,n+\tau-1)}$, and $\mathbf{x}^{(n+K\tau)} = [\boldsymbol{x}^{(n+K\tau)}, \ \boldsymbol{x}^{(n+K\tau)+1}, \ \cdots, \ \boldsymbol{x}^{(n+(K+1)\tau-1)}]$, then we have

$$\mathbf{x}^{(n+(K+1)\tau)} = \mathbf{x}^{(n+K\tau)}\mathscr{P}^{(n)}. \tag{2.131}$$

Under some stability conditions (which have to be determined for each case) we have $\mathbf{x}^{(n+K\tau)}|_{K\to\infty} = \mathbf{x}^{(n)}$. Hence we have

$$\mathbf{x}^{(n)} = \mathbf{x}^{(n)}\mathscr{P}^{(n)}.$$

2.9.9.4 Time-inhomogeneous and spatially-homogeneous QBD

Suppose we have a case where

$$P^{(n)} = \begin{bmatrix} B^{(n)} & C^{(n)} & & & \\ E^{(n)} & A_1^{(n)} & A_0^{(n)} & & \\ & A_2^{(n)} & A_1^{(n)} & A_0^{(n)} & \\ & & \ddots & \ddots & \ddots \end{bmatrix}, \tag{2.132}$$

i.e. the DTMC is time-inhomgeneous but spatially-homogeneous. In this case the matrix $\mathscr{P}^{(n)}$ can be re-blocked. For example, if $\tau = 2$ then

$$\mathscr{P}^{(n)} = P^{(n,n+1)} = \begin{bmatrix} A_{0,0}^{(n,n+1)} & A_{0,1}^{(n,n+1)} & A_{0,2}^{(n,n+1)} & & & \\ A_{1,0}^{(n,n+1)} & A_{1,1}^{(n,n+1)} & A_{1,2}^{(n,n+1)} & A_{1,3}^{(n,n+1)} & & \\ A_{2,0}^{(n,n+1)} & A_{2,1}^{(n,n+1)} & A_{2,2}^{(n,n+1)} & A_{2,3}^{(n,n+1)} & A_{2,4}^{(n,n+1)} & \\ & A_{3,1}^{(n,n+1)} & A_{3,2}^{(n,n+1)} & A_{3,3}^{(n,n+1)} & A_{3,4}^{(n,n+1)} & A_{3,5}^{(n,n+1)} \\ & \ddots & \ddots & \ddots & \ddots & \ddots \end{bmatrix}, \tag{2.133}$$

which results in

$$\mathscr{P}^{(n)} = \begin{bmatrix} B_{0,0}^{(n,n+1)} & B_{0,1}^{(n,n+1)} & C_{0,2}^{(n,n+1)} & & & \\ B_{1,0}^{(n,n+1)} & B_{1,1}^{(n,n+1)} & A_1^{(n,n+1)} & A_0^{(n,n+1)} & & \\ B_{2,0}^{(n,n+1)} & A_3^{(n,n+1)} & A_2^{(n,n+1)} & A_1^{(n,n+1)} & A_0^{(n,n+1)} & \\ & A_4^{(n,n+1)} & A_3^{(n,n+1)} & A_2^{(n,n+1)} & A_1^{(n,n+1)} & A_0^{(n,n+1)} \\ & \ddots & \ddots & \ddots & \ddots & \ddots \end{bmatrix}. \tag{2.134}$$

This can be reblocked to have

$$\mathscr{P}^{(n)} = \begin{bmatrix} \mathscr{B}^{(n)} & \mathscr{C}^{(n)} & & & \\ \mathscr{E}^{(n)} & \mathscr{A}_1^{(n)} & \mathscr{A}_0^{(n)} & & \\ & \mathscr{A}_2^{(n)} & \mathscr{A}_1^{(n)} & \mathscr{A}_0^{(n)} & \\ & & \ddots & \ddots & \ddots \end{bmatrix},$$

where for example

$$\mathscr{A}_1^{(n)} = \begin{bmatrix} A_2^{(n,n+1)} & A_1^{(n,n+1)} \\ A_3^{(n,n+1)} & A_2^{(n,n+1)} \end{bmatrix}.$$

For the general case of $1 \leq \tau < \infty$ we can write and reblock $\mathscr{P}^{(n)}$ into a QBD type transition matrix. Given that $\mathscr{A}^{(n)} = \sum_{k=0}^{2} \mathscr{A}_k^{(n)}$ and let $\boldsymbol{\pi}(n)$ be the stationary distribution associated with it, i.e.

$$\boldsymbol{\pi}(n) = \boldsymbol{\pi}(n)\mathscr{A}^{(n)}, \ \boldsymbol{\pi}(n)\mathbf{1} = 1. \tag{2.135}$$

The stability conditions for standard QBDs apply, i.e. the system is stable iff

$$\boldsymbol{\pi}(n)(\mathscr{A}_1^{(n)} + 2\mathscr{A}_2^{(n)})\mathbf{1} > 1. \tag{2.136}$$

For a stable system we compute the stationary distribution of the transition matrix as

$$\boldsymbol{x}^{(n+j+1)} = \boldsymbol{x}^{(n+j)} R^{(n)}, \tag{2.137}$$

where $R^{(n)}$ is the minimal non-negative solution to the matrix equation

$$R^{(n)} = \mathscr{A}_0^{(n)} + R^{(n)}\mathscr{A}_k^{(n)} + (R^{(n)})^2 \mathscr{A}_k^{(n)}. \tag{2.138}$$

In summary, standard matrix-analytic results can be used for the case of time-inhomogeneous DTMCs that are periodic.

2.10 Software Tools for Matrix-Analytic Methods

There are now software tools available to the public for analysing the M/G/1, GI/M/1 and QBD systems. The website is:

`http://win.ua.ac.be/~vanhoudt/`

The packages use efficient algorithms for the three systems and do provide supporting documents and papers for them also.

Chapter 3
Characterization of Queueing Systems

3.1 Introduction

Queueing systems, by description, are really any types of service systems where customers may have to wait in a queue for service when the server is not available or busy attending to other customers. These types of systems are prevalent in our everyday life activities, ranging from the simple fast food restaurants, transportation systems, computer systems, telecommunication systems, manufacturing systems, etc. Usually there are customers which could be people or items, where items could be data packets, parts in a manufacturing plant, etc. Also in such a system there is or are service providers who provide the facilities for the items to be served.

In order to articulate the subject of this chapter we select an example of a queueing system for discussion. Consider the Internet system. People send packets through this system to be processed and forwarded to a destination. One could consider the packets as the items that need to be processed for a customer. The Internet service provider (ISP) ensures that the customer is provided with the resources to get the packets processed and sent to the correct destination with minimum delay and loss probability. The customer pays for a service and expects a particular Quality of Service (QoS) commensurate with how much she pays. In an utopian world the customer would want the packet processed immediately and the ISP would want to obtain the maximum revenue possible without incurring any costs! In the real world the resources needed to process the packets costs money and the ISP needs to make profit so it wishes to provide the most cost effective amount of resources while the customer who pays for the service sets her QoS target that needs to be met by the ISP.

In designing a queueing system we need to find the optimal configurations and rules that will optimize the profit for the ISP and meet the QoS of customers. In order to do this we need to understand how the queueing system will perform under different configurations and rules. The list of some of the measures of interest to a queueing system operator are as follows:

A.S. Alfa, *Queueing Theory for Telecommunications*,
DOI 10.1007/978-1-4419-7314-6_3, © Springer Science+Business Media, LLC 2010

3.1.0.5 Performance Measures

1. **Queue Length:** The queue length refers to the number of items, in this case packets, that are waiting in some location, or buffer, to be processed. This is often an indication of how well a queueing system is performing. The longer the queue length the worse its performance from a user's point of view, even though this may be argued not to be quite correct all the time.
2. **Loss Probability:** If the buffer space where items have to wait is limited, which often is for real systems, then some items may not be able to wait if they arrive to find the waiting space to be full. We usually assume that such items are lost and may retry by arriving again at a later point in time. In data packet systems the loss of a packet may be very inconvenient and customers are concerned about this loss probability. The higher its value the worse is the system performance from the customer's perspective.
3. **Waiting Times:** waiting time or delay as referred to by some customers is the duration of the time between an item's arrival and when it starts to receive service. This is the most used measure of system performance by customers. Of course the longer this performance measure is the worse is the perception of the system from a customer's point of view.
4. **System Time:** This is simply the waiting time plus the the time to receive service. It is perceived in the same way as the waiting time except when dealing with a preemptive system where an item may have its service interrupted sometimes.
5. **Work Load:** Whenever items are waiting to be processed a server or the system manager is interested in knowing how much time is required to complete processing the items in the system. Of course, if there are no items in the system the workload will be zero. Work load is a measure of how much time is needed to process the waiting items and it is the sum of the remaining service time of the item in service and the sum of the service times of all the waiting items in a work conserving system. In a work conserving system no completed processing is repeated and no work is wasted. A queueing system becomes empty and the server becomes idle once the workload reduces to zero.
6. **Age Process:** In the age process we are interested in how long the item receiving service has been in the system, i.e. the age of the item in process. By age we imply that the item in service was "born" at its arrival time. The age is a measure that is very useful in studying queues where our interest is usually only in the waiting times. This is also very useful in analysing queues with heterogeneous items that are procesed in a first come first served (FCFS) order, also known as first in first out (FIFO) order.
7. **Busy Period:** This is a measure of the time that expires from when a server begins to process after an empty queue to when the queue becomes empty again for the first time. This measure is more of interest to the ISP, who wishes to keep his resources fully utilized. So the higher this value the happier an ISP. However if the resource that is used to provide service is human such as in a bank, grocery store, etc, then there is a limit to how long a service provider wishes to keep a server busy before the server becomes ineffective.

8. **Idle Period:** This is a measure of the time that elapses from when the last item is processed leaving the queue empty to when the service begins again usually after the first item arrival. This is like the "dual" of the busy period. The longer the idle period is the more wasteful are the resources. A service provider wishes to keep the idle period as small as possible.
9. **Departure Times:** This is a measure of the time an item completes service in a single node system. This measure is usually important in cases where an item has a departure time constraint and also where a departing item has to go into another queue for processing. When a single node queue serves an input queue to another queue downstream, the departure process forms the arrival process into the next queue. How this departure process is characterized determines the performance of the downstream queue.

Consider the following process involving 20 messages. For the nth message let us define A_n as its arrival time, S_n as its service time requirement, and D_n as its departure time from the system. We are assuming a FIFO service discipline. The values of A_n and S_n are given in the table below.

n	A_n	S_n
1	0	3
2	2	6
3	4	1
4	6	3
5	7	6
6	12	2
7	20	4
8	21	3
9	29	3
10	31	6
11	32	2
12	38	1
13	41	6
14	43	3
15	44	3
16	48	2
17	49	7
18	52	6
19	54	4
20	58	3

We use this example to illustrate what some of the above mentioned measures really mean. Figs (3.1) to (3.5) show some of the sample paths of this system.

These are just few of the fundamental measures of interest to a queue analyst. There are several others that come up and are specific to some queues and will be discussed as they appear in the book. However, not all of the measures mentioned above will be presented for all queues in the book.

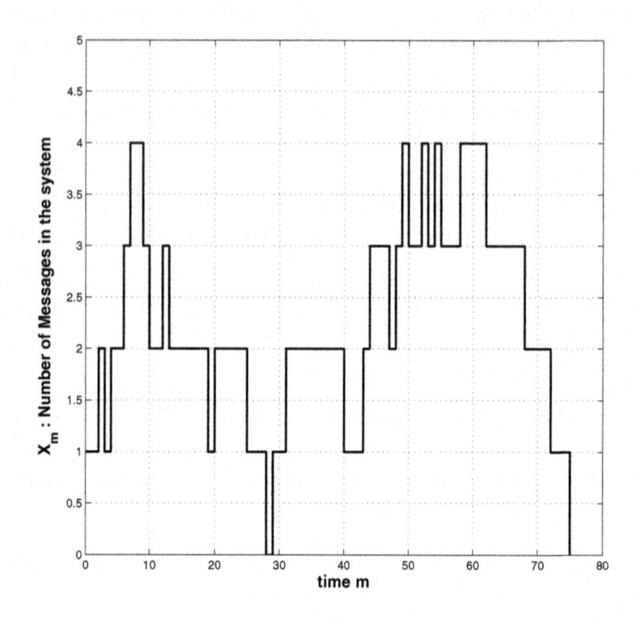

Fig. 3.1 Number in system at arbitrary times

3.1.1 Factors that affect the measures

Many factors determine these performance measures. Some of them are determined by the service provider and other by the customers. The service provider determines what configurations to make the queueing system. For example, the service provider determines:

1. the number of servers to provide– the more the number of servers in parallel the faster will the system be in providing service, but this would involve more cost to the service provider
2. the rule to use for processing items whether it is first come first served (FCFS) (also known as fist in first out (FIFO)), last come first served (LCFS) (also known as LIFO - last in first out), service in random order (SIRO), priority (PR), etc. – this affects how fair the service is to the customers
3. how large the buffer space should be, i.e. the place to store items that are waiting for service when the server is not available – the larger the buffer the less will the loss probability be, but the waiting times for those who have to wait will increase – a trade off.
4. several other aspects

These are all design aspects and can be determined to optimize the operations of a queueing system. Some of the other factors that have major impact on the performance measures discussed earlier are the arrival and service processes. These

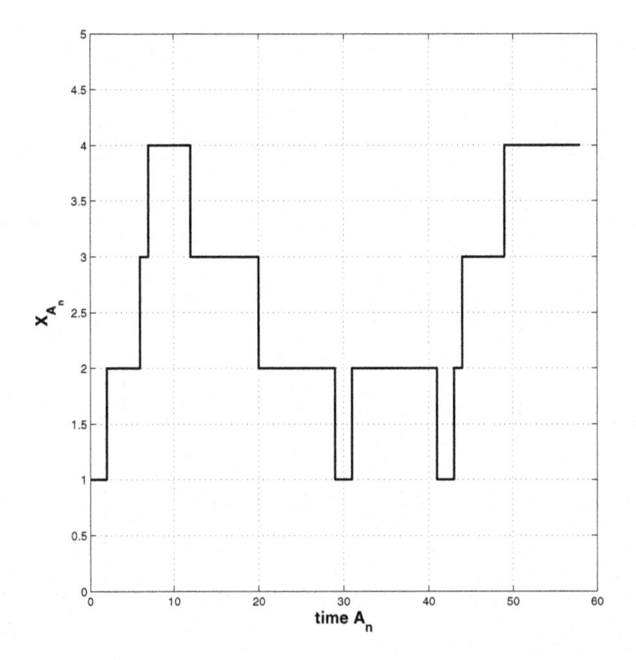

Fig. 3.2 Number in system at arrival times

two quantities are major descriptors of a queueing system and are stochastic by nature. It is simple to expect that high arrival rates would lead to increase in queue lengths, waiting times and busy period. High service rates on the other hand will improve these measures. However, high fluctuation rates of these arrival and service processes could lead to complicated system behaviours. We therefore need to understand these processes very well. The rest of this chapter focuses on the arrival and service processes.

3.2 Arrival and Service Processes

For every queuing system we have to specify both the arrival and the service processes very clearly. They are some of the key characteristics that determine the performance of a queueing system. If the times between arrivals of items are generally short and the service times are long it is clear that the queue will build up. Most important is how those times vary, i.e. the distribution of the interarrival times and service times. We present the arrival and service processes that commonly occur in most queuing systems.

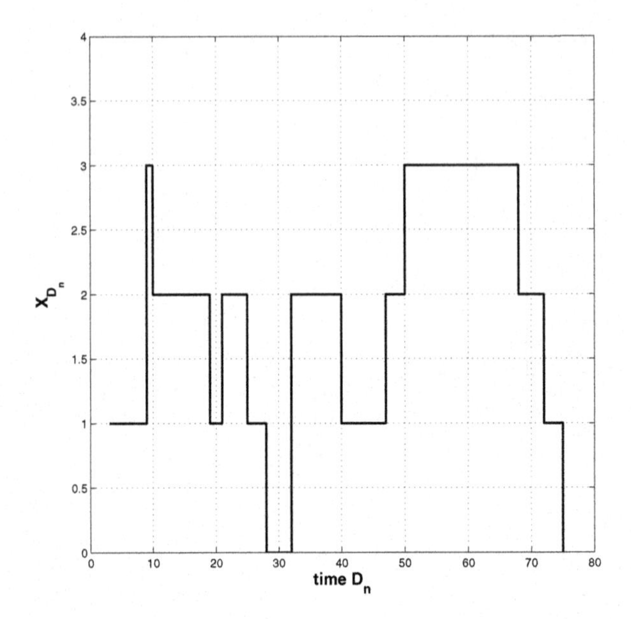

Fig. 3.3 Number in system at departure times

These two processes are usually discussed separately in most literature. But because the distributions used for service processes can also be used for the interarrival times, we combine them in the same presentation.

Consider a situation where packets arrive to a router at discrete time points. Let J_n be the arrival time of the nth packet. Further let A_n be the inter-arrival time between the nth and $(n-1)^{th}$ packet, i.e. $A_n = J_n - J_{n-1}$, $n = 1, 2, \cdots$. We let J_0 be the start of our system and assume an imaginary zeroth packet so that $A_1 = J_1 - J_0$. If $A_1, A_2, A_3, \ldots$ are independent random variables then the time points $J_1, J_2, J_3, \ldots$ are said to form renewal points. If also $A_n = A$, $\forall n$, then we say the interarrival times have identical distributions. We present the distributions commonly assumed by A. We will not be concerned at this point about A_1 since it may have a special behaviour. For all intents and purposes we assume it is a well behaved interval and not a residual. Dealing with the residuals at this point would distract from the main discussion in this chapter.

Let t_n^0 be the start time of the service time of the nth packet and let t_n^f be the completion time of its service time duration, i.e. $S_n = t_n^f - t_n^0$. If $S_n = S$, $\forall n$, then we say that the service times have identical distributions. We can also consider a system that is saturated with a queue of packets needing service. Let C_n be the completion time of the nth packet. Then $S_n = C_n - C_{n-1}$.

First let us briefly introduce renewal process, which is related to some of these arrival and service processes which are independent.

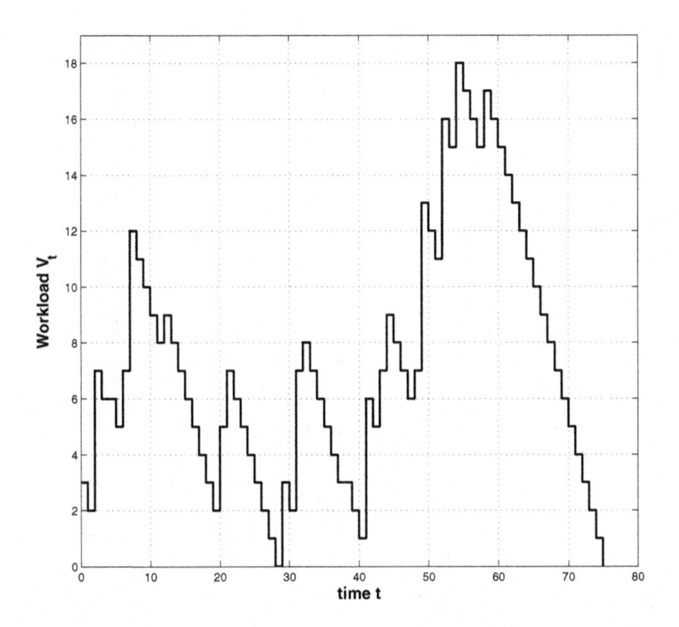

Fig. 3.4 Workload at arbitrary times

3.2.1 Renewal Process

Let X_n be the interevent time between the $(n-1)$th and nth events such as arrivals. In our context X_n could be A_n or S_n. If $Y_0 = 0$ represents the start of our observing the system, then $Y_n = \sum_{i=0}^{n} X_i$ is the time of occurrence of the nth event. For example, $J_n = \sum_{i=0}^{n} A_i$ is the arrival time of the nth customer. We shall assume for simplicity that $X_0 = 0$ and $X_1, X_2, \ldots$ are independent and identically distributed (iid) variables (note that it is more general to assume that X_0 has a value and it has a different distribution than the rest, but that is not very critical for our purpose here).

If we consider the duration of X_n as how long the nth interevent lasts, then Y_n is the point in time when the nth process starts and then lasts for X_n time units. The time Y_n is the regeneration point or the renewal epoch. We assume that $P\{X_n > 0\} = p_0^n > 0$, $\forall n > 1$, and we let $\mathbf{p}_n$ be the collection $\mathbf{p}_n = (p_1^n, p_2^n, \ldots)$, where $p_i^n = Pr\{X_n = i\}$, $\forall n \geq 1$. We define the $z-$transform of this interevent times as

$$p_n^*(z) = \sum_{j=1}^{\infty} p_j^n z^j, \quad |z| \leq 1, \ \forall n. \tag{3.1}$$

Let the mean number of events per unit time be given by μ then the mean interevent time

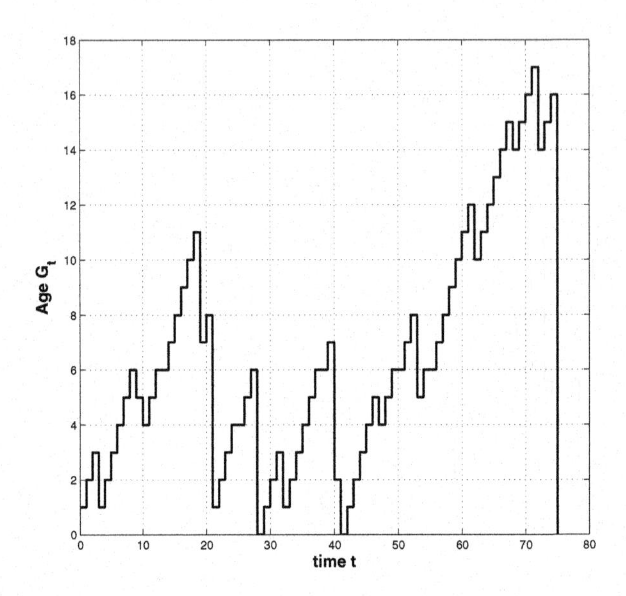

Fig. 3.5 Age at arbitrary times

$$E[X_n] = \frac{dp_n^*(z)}{dz}|_{z \to 1} = \mu^{-1} > 0, \ \forall n. \tag{3.2}$$

Since $X_1, X_2, \ldots$ are independent we can obtain the distribution of Y_n as the convolution sum of $X_1, X_2, \ldots$ as follows

$$\Pr\{Y_n\} = \Pr\{X_1 * X_2 * \ldots * X_n\}. \tag{3.3}$$

Letting $q_j^n = Pr\{Y_n = j\}$ and $q_n^*(z) = \sum_{j=1}^{\infty} q_j^n z^j, \ |z| \leq 1$ we have

$$q_n^*(z) = p_1^*(z) p_2^*(z) \cdots p_n^*(z). \tag{3.4}$$

For example,

$$Pr\{Y_2 = j\} = \sum_{v=1}^{j-1} Pr\{X_1 = v\} Pr\{X_2 = j - v\}.$$

If $X_1 = X_2 = X$, then $p_1^*(z) = p_2^*(z) = p^*(z)$ and hence we have

$$q_2^*(z) = (p^*(z))^2.$$

Further let

$$p_1 = 0.1, \ p_2 = 0.3, \ p_3 = 0.6, \ p_j = 0, j \geq 4,$$

then

$$p^*(z) = 0.1z + 0.3z^2 + 0.6z^3$$

and

$$q_2^*(z) = 0.01z^2 + 0.06z^3 + 0.21z^4 + 0.36z^5 + 0.36z^6.$$

From this we obtain

$$q_1^2 = 0,\ q_2^2 = 0.01,\ q_3^2 = 0.06,\ q_4^2 = 0.21,\ q_5^2 = 0.36,\ q_6^2 = 0.36,\ q_j^2 = 0, \forall j \geq 7.$$

3.2.1.1 Number of renewals

Let Z_n be the number of renewals in $(0, n)$ then

$$Z_n = \max\{m \geq 0 | Y_m \leq n\} \tag{3.5}$$

Hence,

$$P\{Z_n \geq j\} = P\{Y_j \leq n\} = \sum_{v=1}^{n} P\{Y_j = v\} \tag{3.6}$$

If we let $m_n = E[Z_n]$ be the expected number of renewals in $[0, n)$, it is straightforward to show that

$$m_n = \sum_{j=1}^{n} \sum_{v=j}^{n} P\{Y_j = v\} \tag{3.7}$$

m_n is called the *renewal function*. It is a well known elementary renewal theorem that te mean renewal rate μ is given by

$$\mu = \lim|_{n\to\infty} \frac{m_n}{n} \tag{3.8}$$

For a detailed treatment of renewal process, the reader is referred to Wolff [106]. Here we only give a skeletal proof.

In what follows, we present distributions that are commonly used to describe arrival and service processes. Some of them are of the renewal types.

3.3 Special Arrival and Service Processes in Discrete Time

3.3.1 Geometric Distribution

Geometric distribution is the most commonly used discrete time interarrival or service time distribution. Its attraction in queuing theory is its lack of *memory property*.

Consider a random variable X which has only two possible outcomes - *success* or *failure*, represented by the state space $\{0, 1\}$, i.e. 0 is failure and 1 is success. Let q be the probability that an outcome is a failure, i.e. $Pr\{X = 0\} = q$ and $Pr\{X = 1\} = p = 1 - q$ be the probability that it is a success.

The mean number of successes in one trial is given as

$$E[X] = 0 \times q + 1 \times p = p$$

Let $\theta^*(z)$ be the $z-$transform of this random variable and given as

$$\theta^*(z) = q + pz.$$

Also we have

$$E[X] = \frac{d(q+pz)}{dz}|_{z \to 1} = p. \tag{3.9}$$

Suppose we carry out an experiment which has only two possible outcomes 0 or 1 and each experiment is independent of the previous and they all have the same outcomes X, we say this is a Bernoulli process,and $\theta(z)$ is the $z-$transform of this Bernoulli process. Further, let τ be a random variable that represents the time (number of trials) by which the first success occurs, where all the trials are independent of each other. It is simple to show that if the first success occurs at the τth trial then the first $\tau - 1$ trials must have been failures. The random variable τ has a geometric distribution and

$$\Pr\{\tau = n\} = q^{n-1}p, \; n \geq 1. \tag{3.10}$$

The mean interval for a success is

$$E[X] = \sum_{n=1}^{\infty} nq^{n-1}p = p^{-1}. \tag{3.11}$$

We chose to impose the condition that $n \geq 1$ because in the context of discrete time queueing systems our interarrival times and service times have to be at least one unit of time long, respectively. Let the $z-$ transform of this geometric distribution be $T^*(z)$, we have

$$T^*(z) = pz(1 - qz)^{-1}, \; |z| \leq 1. \tag{3.12}$$

We can also obtain the mean time to success from the $z-$transform as

$$E[X] = \frac{T^*(z)}{dz}|_{z \to 1} = p^{-1}. \tag{3.13}$$

In the context of arrival process, T is the interarrival time. Success implies an arrival. Hence, at any time, the probability of an arrival is p and no arrival is q. This is known as the Bernoulli process.

Let Z_m be the number of arrivals in the time interval $[0, m]$, with $Pr\{Z_m = j\} = a_{m,j}$, $j \geq 0$, $m \geq 1$ and $a_m^*(z) = \sum_{j=0}^{m} a_{m,j} z^j$. Since the outcome of time trial follows a Bernoulli process then we have

$$a_m^*(z) = (q + pz)^m, \; m \geq 1, \; |z| \leq 1. \tag{3.14}$$

The distribution of Z_m is given by the Binomial distribution as follows:

$$\Pr\{Z_m = i\} = a_{m,i} = \binom{m}{i} q^{m-i} p^i, \; 0 \le i \le m \tag{3.15}$$

The variable $a_{m,i}$ is simply the coefficient of z^i in the term $a_m^*(z)$. The mean number of successes in m trials is given as

$$E[Z_m] = \frac{a_m^*(z)}{dz}|_{z\to 1} = mp. \tag{3.16}$$

In the context of service times, T is the service time of a customer. Success implies the completion of a service. Hence, at any time the probability of completing an ongoing service is p and no service completion is q.

This is what is known as the lack of memory property.

3.3.1.1 Lack of Memory Property:

This lack of memory property is a feature that makes geometric distribution very appealing for use in discrete stochastic modelling. It is shown as follows:

$$\Pr\{T = n+1 | T > n\} = \frac{\Pr\{T = n+1\}}{\Pr\{T > n\}} = \frac{q^n p}{\sum_{m=n+1}^{\infty} q^{m-1} p} = \frac{q^n p}{q^n} = p \tag{3.17}$$

which implies that the duration of the remaining portion of a service time is independent of how long the service had been going on, i.e. *lack of memory*.

The mean interarrival time or mean service time is given by p^{-1}.

Throughout the rest of this book when we say a distribution is geometric with parameter p we imply that it is governed by a Bernoulli process that has the success probability of p.

While there are several discrete distributions of interest to us in queueing theory, we find that most of them can be studied under the general structure of what is known as the Phase type distribution. So we go right into discussing this distribution.

3.3.2 Phase Type Distribution

Phase type distributions are getting to be very commonly used these days after Neuts [80] made them very popular and easily accessible. They are often referred to as the PH distribution. The PH distribution has become very popular in stochastic modelling because it allows numerical tractability of some difficult problems and in addition several distributions encountered in queueing seem to resemble the PH distribution. In fact, Johnson and Taaffe [60] have shown that most of the commonly occurring distributions can be approximated by the phase type distributions using moment matching approach based on three moments. The approach is based on using mixtures of two Erlang distributions - not necessarily of common order. They seem to obtain very good fit for most attempts which they carried out. Other works of fitting phase-type distributions include those of Asmussen and Nerman [20], Bobbio and Telek [25] and Bobbio and Cumani [24].

Phase type distributions are distributions of the time until absorption in an absorbing DTMC. If after an absorption the chain is restarted, then it represents the distribution of a renewal process.

Consider an (n_t+1) absorbing DTMC with state space $\{0,1,2,\cdots,n_t\}$ and let state 0 be the absorbing state. Let T be an m-dimension substochastic matrix with entries

$$T = \begin{bmatrix} t_{1,1} & \cdots & t_{1,n_t} \\ \vdots & \vdots & \vdots \\ t_{n_t,1} & \cdots & t_{n_t,n_t} \end{bmatrix}$$

and also let

$$\boldsymbol{\alpha} = [\alpha_1, \alpha_2, \ldots, \alpha_{n_t}], \text{ and } \boldsymbol{\alpha}\mathbf{1} \leq 1$$

In this context, α_i is the probability that the system starts from a transient state i, $1 \leq i \leq n_t$, and T_{ij} is the probability of transition from a transient state i to a transient state j. We say the phase type distribution is characterized by $(\boldsymbol{\alpha}, T)$ of dimension n_t. We also define α_0 such that $\boldsymbol{\alpha}\mathbf{1} + \alpha_0 = 1$.

The transition matrix P of this absorbing Markov chain is given as

$$P = \begin{bmatrix} 1 & 0 \\ \mathbf{t} & T \end{bmatrix} \tag{3.18}$$

where $\boldsymbol{t} = \mathbf{1} - T\mathbf{1}$.

The phase type distribution with parameters $\boldsymbol{\alpha}$ and T is usually written as PH distribution with representation $(\boldsymbol{\alpha}, T)$. The matrix T and vector $\boldsymbol{\alpha}$ satisfy the following conditions:

Conditions:

1. Every element of T is between 0 and 1, i.e. $0 \leq T_{i,j} \leq 1$.
2. At least for one row i of T we have $\sum_{j=1}^{n_t} T_{i,j}\mathbf{1} < 1$.
3. The matrix T is irreducible.
4. $\boldsymbol{\alpha}\mathbf{1} \leq 1$ and $\alpha_0 = 1 - \boldsymbol{\alpha}\mathbf{1}$

Throughout this book, whenever we write a PH distribution $(\boldsymbol{\alpha}, T)$ there is always a bolded lower case column vector associated with it; in this case $\mathbf{t}$. As another example, if we have a PH distribution $(\boldsymbol{\beta}, S)$ then there is a column vector $\mathbf{s}$ which is given as $\mathbf{s} = \mathbf{1} - S\mathbf{1}$.

If we now define p_i as the probability that the time to absorption into state n_t+1 is i, then we have

$$p_0 = \alpha_0 \quad (3.19)$$

$$p_i = \alpha T^{i-1}\mathbf{t}, \ \ i \geq 1 \quad (3.20)$$

Let $p^*(z)$ be the $z-$transform of this PH distribution, then

$$p^*(z) = \alpha_0 + z\boldsymbol{\alpha}(I-zT)^{-1}\mathbf{t}, \ |z| \leq 1. \quad (3.21)$$

Then nth factorial moment of the time to absorption is given as

$$\mu_n^{'} = n!\boldsymbol{\alpha}T^{n-1}(I-T)^{-k}\mathbf{1}. \quad (3.22)$$

Specifically the mean time to absorption is

$$\mu_1^{'} = E[X] = \boldsymbol{\alpha}(I-T)^{-1}\mathbf{1}. \quad (3.23)$$

We will show later in the study of the phase renewal process that also

$$\mu_1^{'} = E[X] = \boldsymbol{\pi}(\boldsymbol{\pi}\mathbf{t})^{-1}, \quad (3.24)$$

where

$$\boldsymbol{\pi} = \boldsymbol{\pi}(T+\mathbf{t}\boldsymbol{\alpha}), \ \ \boldsymbol{\pi}\mathbf{1} = 1.$$

Example:
Consider a phase type distribution with representation $(\boldsymbol{\alpha}, T)$ given as

$$T = \begin{bmatrix} 0.1 & 0.2 & 0.05 \\ 0.3 & 0.15 & 0.1 \\ 0.2 & 0.5 & 0.1 \end{bmatrix}, \ \ \boldsymbol{\alpha} = [0.3\ 0.5\ 0.2], \ \alpha_0 = 0.$$

For this $\alpha_0 = 0$ and $\mathbf{t} = [0.65\ 0.45\ 0.2]^T$. We have

$$p_1 = 0.46,\ p_2 = 0.2658,\ p_3 = 0.1346,\ p_4 = 0.0686,\ p_5 = 0.0349,$$

$$p_6 = 0.0178,\ p_7 = 0.009,\ p_8 = 0.0046,\ p_9 = 0.0023,\ p_{10} = 0.0012,$$

$$p_{11} = 0.00060758,\ p_{12} = 0.0003093,\ \cdots$$

Alternatively we may report our results as complement of the cummulative distribution, i.e.

$$P_k = Pr\{X \leq k\} = \boldsymbol{\alpha}T^{k-1}(I-T)^{-1}\mathbf{t}.$$

This is given as

$$P_1 = 1.0,\ P_2 = 0.54,\ P_3 = 0.2743,\ P_4 = 0.1397,\ P_5 = 0.0711,\ P_6 = 0.0362,\ \cdots.$$

3.3.2.1 Two very important closure properties of phase type distributions:

Consider two discrete random variables X and Y that have phase type distributions with representations $(\boldsymbol{\alpha},T)$ and $(\boldsymbol{\beta},S)$.

1. **Sum:** Their sum $Z = X+Y$ has a phase type distribution with representation $(\boldsymbol{\delta},D)$ with
$$D = \begin{bmatrix} T & \mathbf{t}\boldsymbol{\beta} \\ \mathbf{0} & S \end{bmatrix}, \quad \boldsymbol{\delta} = [\boldsymbol{\alpha} \;\; \alpha_0\boldsymbol{\beta}].$$
2. Their minimum $W = min(X,Y)$ has a phase type distribution with representation $(\boldsymbol{\delta},D)$ with
$$D = T \otimes S, \quad \boldsymbol{\delta} = [\boldsymbol{\alpha} \otimes \boldsymbol{\beta}].$$

3.3.2.2 Minimal coefficient of variation of a discrete PH distribution

The coefficient of variation (cv) of a discrete PH distribution has a different behaviour compared to its continuous counterpart. For example, for some integer $K < \infty$ a random variable X with

$$Pr\{X=k\} = \begin{cases} 1, & k=K, \\ 0, & k \neq K, \end{cases} \tag{3.25}$$

can be represented by the discrete PH distribution. This PH distrbution has a cv of zero. This type of case with cv of zero is not encountered in the continuous case. This information can sometimes be used to an advantage when trying to fit a dataset to a discrete PH distribution. In general for the discrete PH the coefficient of variation is a function of its mean.

A general inequality for the minimal cv of a discrete PH distribution was obtained by Telek [96] as follows. Consider the discrete PH distribution $(\boldsymbol{\alpha},T)$ of order n_t. Its mean is given by μ'. Let us write μ' as $\mu' = \lfloor \mu' \rfloor + \langle \mu' \rangle$ where $\lfloor \mu' \rfloor$ is the integer part of μ' and $\langle \mu' \rangle$ is the fractional part with $0 \leq \langle \mu' \rangle < 1$. Telek [96] proved that the cv of this discrete PH distribution written as $cv(\boldsymbol{\alpha},T)$ satisfies the inequality

$$cv(\boldsymbol{\alpha},T) \geq \begin{cases} \frac{\langle \mu' \rangle (1-\langle \mu' \rangle)}{(\mu')^2}, & \mu' < n_t, \\ \frac{1}{n_t} - \frac{1}{\mu'}, & \mu' \geq n_t. \end{cases} \tag{3.26}$$

The proof for this can be found in Telek [96].

Throughout this book we will assume that $\alpha_0 = 0$, since we are dealing with queueing systems for which we do not allow interarrival times or service times to be zero.

3.3.2.3 Examples of special phase type distributions

Some special cases of discrete phase type distributions include:

1. Geometric distribution with $\alpha = 1,\ T = q,\ \mathbf{t} = p = 1 - q$. Then $p_0 = 0$, and $p_i = q^{i-1}p,\ i \geq 1$ and $n_t = 1$. For this distribution, the mean $\mu' = 1/p$ and the $cv(1,q) = 1 - 1/p$

2. Negative binomial distribution with $\boldsymbol{\alpha} = [1,0,0,\ldots]$, and $T = \begin{bmatrix} q & p & & \\ & q & p & \\ & & \ddots & \ddots \\ & & & q \end{bmatrix}$,

   and n_t is the number of successes we are looking for occuring at the ith trial. It is easy to show that

$$p_i = \boldsymbol{\alpha} T^{i-1}\mathbf{t} = \binom{i-1}{n_t - 1} p^{n_t} q^{i-n_t},\ i \geq n_t.$$

3. General discrete distribution with finite support can be represented by a discrete phase type distribution with $\alpha = [1,0,0,\ldots]$ and $0 < t_{ij} \leq 1,\ j = i+1$, and $t_{ij} = 0,\ j \neq i+1$ where n_t is the length of the support. For example, for a constant interarrival time with value of $n_t = 4$ we have $\boldsymbol{\alpha} = [1,\ 0,\ 0,\ 0]$ and

$$T = \begin{bmatrix} 0 & 1 & & \\ 0 & 0 & 1 & \\ 0 & 0 & 0 & 1 \\ 0 & 0 & 0 & 0 \end{bmatrix}.$$

   Note that the general distribution with finite support can also be represented as a phase type distribution with $\boldsymbol{\alpha} = [\alpha_1,\ \alpha_2,\ \cdots,\ \alpha_{n_t}]$ and $t_{ij} = 1,\ i = j-1$, and $t_{ij} = 0,\ i \neq j-1$. For the example of constant interarrival time with value of $n_t = 4$ we have $\boldsymbol{\alpha} = [0,\ 0,\ 0,\ 1]$ and

$$T = \begin{bmatrix} 0 & & & \\ 1 & 0 & & \\ 0 & 1 & 0 & \\ 0 & 0 & 1 & 0 \end{bmatrix}.$$

   The case of general discrete distrubition will be discussed in detail later.

3.3.2.4 Analogy between PH and Geometric distributions

Essentially discrete phase type distribution is simply a matrix version of the geometric distribution. The geometric distribution has parameters p for success (or arrival) and q for failure (no arrival). The discrete phase type with representation $(\boldsymbol{\alpha}, T)$ has $\mathbf{t}\boldsymbol{\alpha}$ for success (arrival) and T for failure (no arrival). So if we consider this in the context of a Bernoulli process we have the $z-$transform of this process as $\theta^*(z) = T + z\mathbf{t}\boldsymbol{\alpha}$. Next we discuss the Phase renewal process.

3.3.2.5 Phase Renewal Process:

Consider an interevent time X which is described by a phase type distribution with the representation $(\boldsymbol{\alpha}, T)$. The matrix T records transitions with no event occurring and the matrix $\mathbf{t}\boldsymbol{\alpha}$ records the occurrence of an event and the re-start of the renewal process. If we consider the interval $(0,n)$ and define the number of renewals in this interval as $N(n)$ and the phase of the PH distribution at phase $J(n)$ at time n, and define

$$P_{i,j}(k,n) = Pr\{N(n) = k, J(n) = j | N(0) = 0, J(0) = i\},$$

and the associated matrix $P(k,n)$ such that $(P(k,n))_{i,j} = P_{i,j}(k,n)$, we have

$$P(0,n+1) = TP(0,n),\ n \geq 0 \tag{3.27}$$

$$P(k,n+1) = TP(k,n) + (\mathbf{t}\boldsymbol{\alpha})P(k-1,n),\ k = 1,2,\cdots; n \geq 0. \tag{3.28}$$

Define

$$P^*(z,n) = \sum_{k=0}^{n} z^k P(k,n),\ n \geq 0.$$

We have

$$P^*(z,n) = (T + z(\mathbf{t}\boldsymbol{\alpha}))^n. \tag{3.29}$$

This is analogous to the $(p+zq)^m$ for the $z-$ transform of the Binomial distribution presented earlier. It is immediately clear that the matrix $T^* = T + \mathbf{t}\boldsymbol{\alpha}$ is a stochastic matrix that represent the transition matrix of the phase process associated with this process.

Secondly, we have T as the matrix analogue of p in the Bernoulli process while $\mathbf{t}\boldsymbol{\alpha}$ is the analogue of q in the Bernoulli process. Hence for one time epoch $T + z(\mathbf{t}\boldsymbol{\alpha})$ is the $z-$transform of an arrival. This phase renewal process can be found in state i in the long run with probability π_i where $\boldsymbol{\pi} = [\pi_1,\ \pi_2,\ \cdots,\ \pi_{n_t}]$ and it is given by

$$\boldsymbol{\pi} = \boldsymbol{\pi}(T + \mathbf{t}\boldsymbol{\alpha}), \tag{3.30}$$

and

$$\boldsymbol{\pi}\mathbf{1} = 1. \tag{3.31}$$

Hence the average number of arrivals in one time unit is

$$E[Z_1] = \boldsymbol{\pi}(0 \times T + 1 \times (\mathbf{t}\boldsymbol{\alpha}))\mathbf{1} = \boldsymbol{\pi}(\mathbf{t}\boldsymbol{\alpha}))\mathbf{1} = \boldsymbol{\pi}\mathbf{t}. \tag{3.32}$$

This is the arrival rate and its inverse is the mean interarrival time of the corresponding phase type distribution, as pointed out earlier on.

Define r_n as the probability of a renewal at time n, then we have

$$r_n = \boldsymbol{\alpha}(T + \mathbf{t}\boldsymbol{\alpha})^{k-1}\mathbf{t}, \quad k \geq 1. \tag{3.33}$$

Let $\boldsymbol{\pi}$ be a solution to the equations

$$\boldsymbol{\pi} = \boldsymbol{\pi}(T + \mathbf{t}\boldsymbol{\alpha}), \quad \boldsymbol{\pi}\mathbf{1} = 1.$$

This $\boldsymbol{\pi}$ represents the probability vector of the PH renewal process being found in a particular phase in the long term. If we observe this phase process at arbitrary times given that it has been running for a while, then the remaining time before an event (or for that matter the elapsed time since an event) also has a phase type distribution with representation $(\boldsymbol{\pi}, T)$. It was shown in Latouche and Ramaswami [67] that

$$\boldsymbol{\pi} = (\boldsymbol{\alpha}(I - T)^{-1}\mathbf{1})^{-1}\boldsymbol{\alpha}(I - T)^{-1}. \tag{3.34}$$

Both the interarrival and service times can be represented by the phase type distributions. Generally, phase type distributions are associated with cases in which the number of phases are finite. However, recently, the case with infinite number of phases has started to receive attention.

3.3.3 The infinite phase distribution (IPH)

The IPH is a discrete phase type distribution with infinite number of phases. It is still represented as $(\boldsymbol{\alpha}, T)$, except that now we have the number of phases $n_t = \infty$. The IPH was introduced by Shi et al [92]. One very important requirement is that the matrix T be irreducible and $T\mathbf{1} \leq 1$, with at least one row being strictly less than 1.

If we now define p_i as the probability that the time to absorption into a state we label as $*$ is i, then we have

$$\begin{aligned} p_0 &= \alpha_0 \\ p_i &= \alpha T^{i-1}\mathbf{t}, \quad i \geq 1 \end{aligned}$$

Every other measure carried out for the PH can be easily derived for the IPH. However, we have to be cautious in many instances. For example, the inverse of $I - T$ may not be unique and as such we have to define it as appropriate for the situation under consideration. Secondly computing p_i above requires special techniques at times depending on the structure of the matrix T. The rectangular iteration was proposed by Shi et al [91] for such computations.

3.3.4 General Inter-event Times

General types of distributions, other than of the phase types, can be used to describe both interarrival and service times. In continuous times it is well known that general distributions encountered in queueing systems can be approximated by continuous time PH distributions. This is also true for discrete distributions. However, discrete distributions have an added advantage in that if the distribution has a finite support then it can be represented exactly by discrete PH. We proceed to show how this is true by using a general intervent time X with finite support and a general distribution given as

$$Pr\{X=j\}=a_j,\ j=1,2,\cdots,n_t<\infty.$$

There are at least two exact PH representations for this distribution, one based on remaining time and the other on elapsed time.

3.3.4.1 Remaining Time Representation

Consider a PH distribution $(\boldsymbol{\alpha},T)$ of dimension n_t. Let

$$\boldsymbol{\alpha}=[a_1,\ a_2,\ \cdots,\ a_{n_t}], \tag{3.35}$$

and

$$T_{i,j}=\begin{cases}1, & j=i-1\\ 0, & \text{otherwise,}\end{cases} \tag{3.36}$$

Then the distribution of this PH is given as

$$\boldsymbol{\alpha}T^{k-1}\mathbf{t}=a_k;\ k=1,2,\cdots,n_t. \tag{3.37}$$

For detailed discussion on this see Alfa [6].

In general, even if the support is not finite we can represent the general interevent times with the IPH, by letting $n_t\to\infty$. For example, consider the interevent times $\mathscr{A}$ which assume values in the set $\{1,2,3,\cdots\}$ with $a_i=Pr\{\mathscr{A}=i\},\ i=1,2,3,\cdots$. It is easy to represent this distribution as an IPH with $\boldsymbol{\alpha}=[a_1,\ a_2,\ \cdots]$ and $T=\begin{bmatrix}0 & 0\\ I_\infty & 0\end{bmatrix}$.

3.3.4.2 Elapsed Time Representation

Consider a PH distribution $(\boldsymbol{\alpha},T)$ of dimension n_t. Let

$$\boldsymbol{\alpha}=[1,\ 0,\ \cdots,\ 0], \tag{3.38}$$

and

$$T_{i,j}=\begin{cases}\tilde{a}_i, & j=i+1\\ 0, & \text{otherwise,}\end{cases} \tag{3.39}$$

where

$$\tilde{a}_i = \frac{u_i}{u_{i-1}}, \; u_i = 1 - \sum_{v=1}^{i} a_v, \; u_0 = 1, \; \tilde{a}_{n_t} = 0.$$

Then the distribution of this PH is given as

$$\alpha T^{k-1}\mathbf{t} = a_k; \; k = 1, 2, \cdots, n_t. \tag{3.40}$$

For detailed discussion on this see Alfa [6]. Similarly we can use IPH to represent this distribution using elapsed time by allowing $n_t \to \infty$.

3.3.5 Markovian Arrival Process

All the interevent times discussed so far are of the renewal types and are assumed to be independent and identically distributed *iid*. However, in telecommunication queueing systems and most other traffic queueing systems for that matter, interarrival times are usually correlated. So the assumption of independent interarrival times is not valid in some instances.

Recently Neuts [79, 82] and Lucantoni [72] presented the Markovain arrival process (MAP) which can handle correlated arrivals and is also tractable mathematically. In what follows, we first describe the single arrival MAP and then briefly present the batch MAP.

Consider an $(n_t + 1)$ state absorbing Markov chain with state space $\{0, 1, 2, \cdots, n_t\}$ in which state 0 is the absorbing state. Define two sub-stochastic matrices D_0 and D_1, both of the dimensions n. The elements $(D_0)_{ij}$ refer to transition from state i to state j without an (event) arrival because the transitions are all within the n_t transient states. The elements $(D_1)_{ij}$ refer to transition from state i into the absorbing state 0 with an instantaneous restart from the transient state j with an (event) arrival during the absorption. We note that the phase from which an absorption occured and the one from which the next process starts are connected and hence this captures the correlation between interarrival times. The matrix $D = D_0 + D_1$ is a stochastic matrix, and we assume it is irreducible. Note that $D\mathbf{1} = \mathbf{1}$. If we define $\{(N_n, J_n), n \geq 0\}$ as the total number of arrivals and the phase of the MAP at time n, then the transition matrix representing this system is

$$P = \begin{bmatrix} D_0 & D_1 & & & \\ & D_0 & D_1 & & \\ & & D_0 & D_1 & \\ & & & \ddots & \ddots \end{bmatrix}. \tag{3.41}$$

Consider the discrete-time Markov renewal process embedded at the arrival epochs and with transition probabilities defined by the sequence of matrices

$$Q(k) = [D_0]^{k-1} D_1, \; k \geq 1. \tag{3.42}$$

The MAP is a discrete-time point process generated by the transition epochs of that Markov renewal process.

Once more let N_m be the number of arrivals at time epochs $1,2,\ldots,m$, and J_m the state of the Markov process at time m. Let $P_{r,s}(n,m) = \Pr\{N_m = n,\ J_m = s \mid N_0 = 0,\ J_0 = r\}$ be the (r,s) entry of a matrix $P(n,m)$. The matrices $P(n,m)$ satisfy the following discrete Chapman-Kolmogorov difference equations:

$$P(n,m+1) = P(n,m)D_0 + P(n-1,m)D_1,\ n \geq 1,\ m \geq 0 \tag{3.43}$$

$$P(0,m+1) = P(0,m)D_0 \tag{3.44}$$

$$P(0,0) = I \tag{3.45}$$

where I is the identity matrix and $P(u,v) = 0$ for $u \geq v+1$.

The matrix generating function

$$P^*(z,m) = \sum_{n=0}^{m} P(n,m)z^n,\ |z| \leq 1 \tag{3.46}$$

is given by

$$P^*(z,m) = (D_0 + zD_1)^m,\ m \geq 0 \tag{3.47}$$

If the stationary vector π of the Markov chain described by D satisfies the equation

$$\boldsymbol{\pi} D = \pi, \tag{3.48}$$

and

$$\boldsymbol{\pi}\mathbf{1} = 1 \tag{3.49}$$

then $\lambda' = \boldsymbol{\pi} D_1 \mathbf{1}$ is the probability that, in the stationary version of the arrival process, there is an arrival at an arbitrary point. The parameter λ' is the expected number of arrivals at an arbitrary time epoch or the discrete arrival rate of the MAP.

It is clear that its kth moments about zero are all the same for $k \geq 1$, because $1^j \boldsymbol{\pi} D_1 \mathbf{1} = \lambda',\ \forall j \geq 1$, since its jth moment about zero is given as $\sum_{k=0}^{\infty} k^j \boldsymbol{\pi} D_k \mathbf{1}$. Hence its variance σ_X^2 is given as

$$\sigma_X = \lambda'' - (\lambda')^2 = \lambda' - (\lambda')^2 = \lambda'(1-\lambda'). \tag{3.50}$$

Its jth autocorrelation factor for the number of arrivals, $ACF(j)$, is given as

$$ACF(j) = \frac{1}{\sigma_X^2}[\boldsymbol{\pi} D_1 D^{j-1} D_1 \mathbf{1} - (\lambda')^2],\ j \geq 1. \tag{3.51}$$

The autocorrelation between interarrival times is captured as follows. Let X_i be the ith interarrival time then we can say that

$$Pr\{X_i = k\} = \boldsymbol{\pi} D_0^{k-1} D_1 \mathbf{1},\ k \geq 1, \tag{3.52}$$

where $\boldsymbol{\pi} = \boldsymbol{\pi}(I - D_0)^{-1}D_1, \ \boldsymbol{\pi}\mathbf{1} = 1$. Then $E[X_i]$, the expected value of $X_i, \ \forall i$, is given as

$$E[X_i] = \sum_{j=1}^{\infty} j\boldsymbol{\pi} D_0^{j-1} D_1 \mathbf{1} = \boldsymbol{\pi}(I - D_0)^{-1}\mathbf{1}. \tag{3.53}$$

The autocorrelation sequence $r_k, \ k = 1, 2, \cdots$ for the interarrival times is thus given as

$$r_k = \frac{E[X_\ell X_k] - E[X_\ell]E[X_k]}{E[X_\ell^2] - (E[X_\ell])^2}, \quad \ell = 0, 1, 2, \cdots. \tag{3.54}$$

Special Cases:

The simplest MAP is the Bernoulli process with $D_1 = q$ and $D_0 = p = 1 - q$. The discrete phase type distribution is also a MAP with $D_1 = \mathbf{t}\boldsymbol{\alpha}$ and $D_0 = T$.

Another interesting example of MAP is the Markov modulated Bernoulli process (MMBP) which is controlled by a Markov chain which has a transition matrix P and rates given by the $\theta = diag(\theta_1, \theta_1, \ldots, \theta_{n_t})$. In this case, $D_0 = (I - \theta)P$ and $D_1 = \theta P$.

Several examples of discrete MAP can be found in Alfa and Neuts [10], Alfa and Chakravarthy [7], Alfa, Dolhun and Chakravarthy [8], Liu and Neuts [71], Park, et al. [86] and Blondia [23].

3.3.5.1 Platoon Arrival Process (PAP)

The PAP is a special case of MAP which occurs in several traffic situations mainly in telecommunications and vehicular types of traffic. It captures distinctly, in addition to correlation in arrival process, the bursts in traffic arrival process termed platoons here.

The PAP is an arrival process with two regimes of traffic. There is a platoon of traffic (group of packets) which has an intraplatoon intervals of arrival times that are identical. The number of arrivals in a platoon is random. At the end of a platoon there is an interplatoon interval between the end of one platoon and the start of the next platoon arrivals. The interplatoon intervals are different from the intraplatoon intervals. A good example is if one observes the departure of packets from a router queue, one observes a platoon departure that consists of traffic departing as part of a busy period. The interplatoon times are essentially the service times. The last packet that departs at the end of a busy period marks the end of a platoon. The next departure will be the head of a new platoon, and the interval between the last packet of a platoon and the first packet of the next platoon is the interplatoon time which in this case is the sum of the service time and the residual interarrival time of a packet into the router. If we consider this departure from the router as an input to another queueing system, then the arrival to this other queueing system is a PAP.

A PAP is described as follows. Let packets arrive in platoons of random sizes with probability mass function *pmf* given as $\{p_k, \ k \geq 1\}$, i.e. p_k is the probability that the number in a platoon is k. Time intervals between arrivals of packets in the same platoon denoted as intraplatoon interarrival times have a *pmf* denoted as $\{p_1(j),\ j \geq 1\}$. The time intervals between the arrival of the last packet in a platoon and the first packet of the next platoon is referred to as the interplatoon interarrival times and has the *pmf* denoted by $\{p_2(j),\ j \geq 1\}$. Let the size of platoon be distributed according to the PH distribution $(\boldsymbol{\delta}, F)$. For this PH let us allow $\delta_0 = 1 - \boldsymbol{\delta}\mathbf{1} \geq 0$, and define $\mathbf{f} = \mathbf{1} - D\mathbf{1}$. Here we have

$$p_k = \begin{cases} \delta_0,\ k = 1 \\ \boldsymbol{\delta} F^{k-1}\mathbf{f},\ k = 1, 2, \cdots \end{cases}. \tag{3.55}$$

The platooned arrival process (PAP) is a discrete-time Markov renewal process whose transition probabilities are described by the sequence of matrices

$$f(j) = \begin{bmatrix} \delta_0 p_2(j) & \boldsymbol{\delta} p_2(j) \\ \mathbf{f} p_1(j) & F p_1(j) \end{bmatrix},\ j \geq 1. \tag{3.56}$$

If we let the intraplatoon and interplatoon interarrival times assume PH distribution then we actually end up with a PAP that is a special case of a MAP. Let $(\boldsymbol{\alpha}_1, T_1)$ be a PH distribution that describes intraplatoon times and $(\boldsymbol{\alpha}_2, T_2)$ a PH distribution describing the interplatoon times. Then the PAP is described by two matrices D_0 and D_1, with

$$D_0 = \begin{bmatrix} T_2 & \mathbf{0} \\ \mathbf{0} & I \otimes T_1 \end{bmatrix}, \tag{3.57}$$

and

$$D_1 = \begin{bmatrix} \delta_0 \mathbf{t}_2 \boldsymbol{\alpha}_2 & \boldsymbol{\delta} \otimes \mathbf{t}_2 \boldsymbol{\alpha}_1 \\ \mathbf{f} \otimes \mathbf{t}_1 \boldsymbol{\alpha}_2 & F \otimes \mathbf{t}_1 \boldsymbol{\alpha}_1 \end{bmatrix}. \tag{3.58}$$

Note that $\mathbf{f} = \mathbf{1} - F\mathbf{1},\ \mathbf{t}_k = \mathbf{1} - T_k\mathbf{1},\ k = 1, 2.$

Next we explain the elements of the marices D_0 and D_1.

For the matrix D_0 we have

- the matrix T_2 captures the phase transitions during an intraplatoon arrival time, and
- the matrix $I \otimes T_1$ captures the transitions during an interarrival time within a platoon.

For the matrix D_1 we have

- the matrix $\delta_0 \mathbf{t}_2 \boldsymbol{\alpha}_2$ captures the arrival of a platoon of a single packet type, with an end to the arrival of a platoon and the initiation of an interplatoon interarrival time process, whilst
- the matrix $\boldsymbol{\delta} \otimes \mathbf{t}_2 \boldsymbol{\alpha}_1$ captures the arrival of a platoon consisting of at least two packets, with an end to the interplatoon interarrival and the initiation of an intraplatoon interarrival time process.
- The matrix $F \otimes \mathbf{t}_1 \boldsymbol{\alpha}_1$ captures the arrival of an intraplatoon packet, with an end to the intraplatoon interarrival and the initiation of a new intraplatoon interarrival time process, whilst

- the matrix $\mathbf{f} \otimes \mathbf{t}_1 \boldsymbol{\alpha}_2$ captures the arrival of the last packet in a platoon, with an end to the intraplatoon interarrival and the initiation of an interplatoon interarrival time process.

A simple example of this is where intraplatoon and interplatoon interarrival times follow geomeric distributions, with $p_1(j) = (1-a_1)^{j-1}a_1,\ p_2(j) = (1-a_2)^{j-1}a_2,\ \ j \geq 1$, and the distribution of platoon size is geometric with $p_0 = 1-b,\ p_k = b(1-c)^{k-1}c,\ k \geq 1$, where $0 < (a_1, a_2, b, c) < 1$. In this case we have

$$D_0 = \begin{bmatrix} 1-a_2 & 0 \\ 0 & 1-a_1 \end{bmatrix}, \quad D_1 = \begin{bmatrix} ba_2 & (1-b)a_2 \\ ba_1 & (1-b)a_1 \end{bmatrix}.$$

For a detail discussion of the discrete PAP see Alfa and Neuts [10].

3.3.5.2 Batch Markovian Arrival Process (BMAP)

Define substochastic matrices $D_k,\ k \geq 0$, such that $D = \sum_{k=0}^{\infty} D_k$ is stochastic. The elements $(D_k)_{ij}$ refer to a transition from state i to state j with $k \geq 0$ arrivals.

If we define π such that

$$\boldsymbol{\pi} D = \boldsymbol{\pi}, \quad \boldsymbol{\pi} \mathbf{1} = 1$$

then the arrival rate $\lambda' = E[X] = \boldsymbol{\pi} \sum_{k=1}^{\infty} k D_k \mathbf{1}$. Let $\lambda'' = E[X^2] = \boldsymbol{\pi} \sum_{k=1}^{\infty} k^2 D_k \mathbf{1}$ be its second moment about zero, then its variance σ_X^2 is given as

$$\sigma_X^2 = \lambda'' - (\lambda')^2. \tag{3.59}$$

Its jth autocorrelation factor $ACF(j)$ is given as

$$ACF(j) = \frac{1}{\sigma_X^2}[\boldsymbol{\pi}(\sum_{k=1}^{\infty} kD_k)D^{j-1}(\sum_{k=1}^{\infty} kD_k)\mathbf{1} - (\lambda')^2], \ \ j \geq 1. \tag{3.60}$$

3.3.6 Marked Markovian Arrival Process

Another class of arrival process of interest in telecommunication is the marked Markovian arrival process (MMAP[K]), with K classes. It is represented by $K+1$ matrices $D_0,\ D_1,\ \cdots,\ D_K$ all of dimension $n_t \times n_t$. The elements $(D_k)_{i,j},\ k = 1, 2, \cdots, K$, represent transitions from state i to state j with type k packet arrival and $(D_0)_{i,j}$ represents no arrival. Let $D = \sum_{v=0}^{K} D_v$, then D is a stochastic matrix. For a detailed discussion of this class of arrival process see He and Neuts [53].

3.3.7 Semi Markov Processes

Service times with correlations can be described by a class of semi-Markov processes, which are the analogue of the MAP. For such examples, see Alfa and Chakravarthy [7] and Lucantoni and Neuts [74].

3.3.8 Data Fitting for PH and MAP

Fitting of PH distribution is both a science and an art. This is because PH representation of a given probability distribution function is not unique. As a trivial example, the geometric distribution $r_k = q^{k-1}p,\ k = 1,2,\cdots$, can be represented as a PH distribution with parameters $(1,q)$ or even as $(\boldsymbol{\alpha},T)$, where $\boldsymbol{\alpha} = [0,\ 1,\ 0]$ and $T = \begin{bmatrix} b_1 & b_2 & b_3 \\ 0 & q & 0 \\ c_1 & c_2 & c_3 \end{bmatrix}$, where $0 \leq (b_i, c_i) \leq 1$ and $b_1 + b_2 + b_3 \leq 1,\ c_1 + c_2 + c_3 \leq 1$. Several examples of this form can be presented. It is also known that PH representation is non-minimal, as demonstrated by the last example. So, fitting a PH distribution usually involves selecting the number of phases in advance and then finding the best fit using standard statistical methods such as the maximum likelihood method of moments. Alternatively one may select the structure of the PH and then find the best fitting order and parameters. So the art is in the pre-selection process which is often guided by what the PH distribution is going to be used for in the end. For example if the PH is for representing service times in a queueing problem we want to have a small dimension so as to reduce the computational load associated with the queueing model. In some other instances the structure of the PH may be more important if a specific structure will make computation easier.

In general assume we are given a set of N observations of interevent times $y_1,\ y_2,\ \cdots,\ y_N$, with $\mathbf{y} = [y_1,\ y_2,\ \cdots,\ y_N]$. If we want to fit a PH distribution $(\boldsymbol{\alpha},T)$ of order n and/or known structure to this set of observations we can proceed by of the following two methods:

- **Method of moments:** We need to estimate a maximum of $n^2 + n - 1$ parameters. This is because we need a maximum of n^2 parameters for the matrix T and $n-1$ parameters for the vector $\boldsymbol{\alpha}$. If we have a specific structure in mind then the number of unknowns that need to be determined can be reduced. For example if we want to fit a negative binomial distribution of order n then all we need is one parameter. But if it is a general negative binomial then we need n parameters. Let the number of parameters needed to be determined by $m \leq n_n^2 - 1$. Then we need to compute m moments of the dataset $\mathbf{y}$, which we write as $\tilde{\mu}_k,\ k = 1,2,\cdots,m$. With this knowledge and also knowing that the factorial moments of the PH distribution are given as $\mu_k^{'} = k!\boldsymbol{\alpha}(I-T)^{-k}\mathbf{1}$ we can then obtain the moments of the PH μ_k from the factorial, equate them to the moments from the observed data. This will lead to a set of m non-linear equations, which need to be solved to obtain the best fitting parameters.

- **Maximum likelihood (ML) method:** This second approach is more popularly used. It is based on the likelihoodness of observing the dataset. Let $f_i(y_i, \boldsymbol{\alpha}, T) = \boldsymbol{\alpha} T^{y_i - 1}\mathbf{t}$, $i = 1, 2, \cdots, N$. Then the likelihood function is

$$\mathscr{L}(\mathbf{y}, \boldsymbol{\alpha}, T) = \prod_{i=1}^{N} f_i(y_i, \boldsymbol{\alpha}, T).$$

  This function is usually converted to its log form and then different methods can be used to find the best fitting parameters.

Data fitting of the PH distribution is outside the scope of this book. However, several methods can be found in Telek [96], Bobbio [26]. Bobbio [26] specifically presented methods for fitting acyclic PH. Acyclic PH is a special PH $(\boldsymbol{\alpha}, T)$ for which $T_{i,j} = 0, \forall i \geq j$ and $\alpha_1 = 1$.

There has not been much work with regards to the fitting of discrete MAP. However, Breuer and Alfa [27] did present the ME algorithm based on ML for estimating parameters for the PAP.

Chapter 4
Single Node Queuing Models

4.1 Introduction

A single node queue is a system in which a customer comes for service only at one node. The service may involve feedback into the same queue. When it finally completes service, it leaves the system. If there is more than one server then they must all be in parallel. It is also possible to have single node queues in which we have several parallel queues and a server (or several servers) move around from queue to queue to process the items waiting; an example is a polling system to be discussed later. Polling systems are used extensively to model medium access control (MAC) in telecommunications. However, when we have a system where an item arrives for service and after completing the service at a location the item joins another queue in a different location for processing. These types of systems are not single node queues. They could be network of queues or queues in tandem. Single node queues are special cases of these systems. Figs (1.1) to (1.4) are single node queues and Figs(1.5) and (1.6) are not.

A single node queue is characterized by the arrival process (A), the service process (B), the number of servers (C) in parallel, the service discipline (D), the available buffer space (E) and the size of the population source (F). Note that the notation (E) is used here to represent the buffer size, whereas in some papers and books it is used to represent the buffer plus the available number of servers. Using Kendall's notations, a single node queue can be represented as A/B/C/D/E/F. For example, if the arrival process is Geometric, service is Phase type, with 2 servers in parallel, First In First Out service discipline, K buffer spaces, and infinite size of the population source, then the queue is represented as Geo/PH/2/FIFO/K/∞ or Geo/PH/2/FCFS/K/∞, since FIFO and FCFS can be used interchangeably. Usually only the first three descriptors are used when the remaining are common occurrence or their omission will not cause any ambiguity. For example, Geo/PH/2 implies that we have geometric interarrival times, Phase type services, two servers in parallel, first in first out service discipline, infinite waiting space and infinite size of population source. When we need to include a finite buffer space the we use the notation

A.S. Alfa, *Queueing Theory for Telecommunications*,
DOI 10.1007/978-1-4419-7314-6_4, © Springer Science+Business Media, LLC 2010

A/B/C/E. For example Geo/Geo/3/K implies geometric interarrival times, geometric service times, three identical serves in parallel and a buffer of size K. For more complex single node queues such as polling systems, feedback systems, etc. we may simply add some texts to give further descriptions.

For all the queues which we plan to study in this chapter, we will assume that the interarrival times and service times are independent of one another. Also, unless otherwise stated, we will assume that the types of service disciplines we are dealing with are FIFO in most of this chapter and the next one when studying the waiting times.

A major emphasis of this book is to present a queuing model analysis via the use of Markov chain techniques, without necessarily overlooking the other popular and effective methods. We therefore use the matrix method very frequently to present this. The aim in setting up every queuing model is to reduce it to a Markov chain analysis, otherwise it would be difficult to solve. Use of recursion is also paramount in presenting the data. The main performance measures that we will focus on are: the number of items in the system, the queue length, and the waiting times. We will present busy period in some instances and the other measures such as workload, departure times and age will only be considered in a few cases of interest. Specifically, workload and age process will be considered for Geo/Geo/1 and PH/PH/1 systems, and departure process for Geo/Geo/1/K and PH/PH/1/K systems.

Considerable effort will be focused in studying the Geo/Geo/1 and the PH/PH/1 systems. The latter is the matrix analogue of the former and a good knowledge of these two classes of systems will go a long way to helping in understanding the general single node queueing systems. Generally the structure of the PH/PH/1 system appears frequently in the book and its results can be extended to complex single node cases with little creative use of the Markov chain idea. Understanding Geo/Geo/1 makes understanding PH/PH/1 simple and besides the former is a special case of the latter.

4.2 Birth-and-Death Process

Birth-and-death (BD) processes are processes in which a single birth can occur at anytime and the rate of birth depends on the number in the system. Similarly, a single death can occur at anytime and the death rate depends on the number in the system. It will be noticed later that a birth-and-death process can only increase the population or decrease it by at most, one unit at a time, hence, its Markov chain is of the tri-diagonal type. The birth-and-death process is very fundamental in queuing theory. The discrete time BD process has not received much attention in the literature. This is partly because it does not have the simple structure that makes it very general for a class of queues as its continuous time counterpart.

In the discrete time version, we assume that the birth-process is state dependent Bernoulli process and the death process follows a state dependent geometric distribution. In what follows, we develop the model more explicitly.

4.3 Discrete time B-D Process

Let a_i = Probability that a birth occurs when there are $i \geq 0$ customers in the system with $\bar{a}_i = 1 - a_i$. Also, let b_i = Probability that a death occurs when there are $i \geq 1$ customers in the systems, with $\bar{b}_i = 1 - b_i$. We let $b_0 = 0$.

If we now define a Markov chain with state space $\{i, i \geq 0\}$, where i is the number of customers in the system, then the transition matrix P of this chain is given as

$$P = \begin{bmatrix} \bar{a}_0 & a_0 & & & \\ \bar{a}_1 b_1 & \bar{a}_1 \bar{b}_1 + a_1 b_1 & a_1 \bar{b}_1 & & \\ & \bar{a}_2 b_2 & \bar{a}_2 \bar{b}_2 + a_2 b_2 & a_2 \bar{b}_2 & \\ & & \ddots & \ddots & \ddots \end{bmatrix} \tag{4.1}$$

If we define $x_i^{(n)}$ as the probability that there are i customers in the system at time n, and letting $\boldsymbol{x}^{(n)} = \left[x_0^{(n)}, x_1^{(n)}, \ldots\right]$ then we have

$$\boldsymbol{x}^{(n+1)} = \boldsymbol{x}^{(n)} P, \text{ or } \boldsymbol{x}^{(n+1)} = \boldsymbol{x}^{(0)} P^{n+1}.. \tag{4.2}$$

If P is irreducible and positive recurrent, then there exists an invariant vector, which is also equivalent to the limiting distribution, $\boldsymbol{x} = \boldsymbol{x}^{(n)}|_{n \to \infty}$ and given as

$$\boldsymbol{x} P = \boldsymbol{x}, \quad \mathbf{1} = 1. \tag{4.3}$$

If we let $\boldsymbol{x} = [x_0, x_1, x_2, \ldots]$, then in an expanded form we can write

$$x_0 = x_0 \bar{a}_0 + x_1 \bar{a}_1 b_1 \tag{4.4}$$

$$x_1 = x_0 a_0 + x_1 \left(\bar{a}_1 \bar{b}_1 + a_1 b_1\right) + x_2 \bar{a}_2 b_2 \tag{4.5}$$

$$x_i = x_{i-1} a_{i-1} \bar{b}_{i-1} + x_i \left(\bar{a}_i \bar{b}_i + a_i b_i\right) + x_{i+1} \bar{a}_{i+1} b_{i+1}, \; i \geq 2 \tag{4.6}$$

This results in

$$x_1 = \frac{a_0}{\bar{a}_1 b_1} x_0 \tag{4.7}$$

from the first equation, and

$$x_i = \prod_{v=0}^{i-1} \frac{a_v \bar{b}_v}{\bar{a}_{v+1} b_{v+1}} x_0, \quad i \geq 1 \tag{4.8}$$

from the other equations, where $b_0 = 0$.

Using the normalizing condition that $\boldsymbol{x}\mathbf{1} = 1$, we obtain

$$x_0 = \left[1 + \sum_{i=1}^{\infty} \prod_{v=0}^{i-1} \frac{a_v \bar{b}_v}{\bar{a}_{v+1} b_{v+1}}\right]^{-1}. \tag{4.9}$$

Note that for stability, we require that $x_0 = \sum_{i=1}^{\infty} \prod_{v=0}^{i-1} (a_v \bar{b}_v)[\bar{a}_{v+1} b_{v+1}]^{-1} < \infty$.

The BD process is a special case of the level dependent QBD discussed in Sections 2.9.7.9 and 2.9.8.1. Here we have scalar entries in the transition matrix instead of block matrices. It is interesting to note that even for this scalar case we do not have simple general solutions. We have to deal with each case as it arises. An example of a case that we can handle is where

$$a_i = a, \text{ and } b_i = b, \ \forall i \geq K < \infty.$$

In this case we have

$$x_i = \prod_{v=0}^{i-1} \frac{a_v \bar{b}_v}{\bar{a}_{v+1} b_{v+1}} x_0, \ \ 1 \leq i \geq K, \tag{4.10}$$

and

$$x_{i+1} = x_i \frac{a\bar{b}}{b\bar{a}}, \ \ i \geq K. \tag{4.11}$$

The normalization equation is applied appropriately.

Later we show that the Geo/Geo/1 and the Geo/Geo/1/K are special cases of this BD process. In the continuous time case it is possible to show that the M/M/s is a special case of the continuous time BD process. However this is not the case for the discrete time BD; the Geo/Geo/s can not be shown to be a special case of the discrete BD process. We discuss this later in the next chapter.

In what follows, we present special cases of the discrete birth-and-death process at steady state.

4.4 Geo/Geo/1 Queues

We consider the case when $a_i = a$, $\forall i \geq 0$ and $b_i = b$, $\forall i \geq 1$. We define the transforms $b^*(z) = 1 - b + zb$ and $a^*(z) = 1 - a + za$. In this case, the transition matrix P becomes

$$P = \begin{bmatrix} \bar{a} & a & & \\ \bar{a}b & \bar{a}\bar{b} + ab & a\bar{b} & \\ & \bar{a}b & \bar{a}\bar{b} + ab & a\bar{b} \\ & & \ddots & \ddots & \ddots \end{bmatrix}. \tag{4.12}$$

We can use several methods to analyze this Markov chain. First, we present the standard algebraic approach, the z-transform approach and then use the matrix-geometric approach.

An example application of this queueing model in telecommunication is where we have packets arriving at a single server (a simple type of router) according to the Bernoulli process. Each packet has a random size with a geometric distribution, hence the time it takes to process each packet follows a geometric distribution. Because the packets arrive according to the Bernoulli process then their interarrival times follow the geometric distribution. Hence we have a Geo/Geo/1 queueing system.

4.4.1 Algebraic Approach

$$x_0 = x_0\bar{a} + x_1\bar{a}b \tag{4.13}$$

$$x_1 = x_0 a + x_1\left(\bar{a}\bar{b} + ab\right) + x_2\bar{a}b \tag{4.14}$$

$$x_i = x_{i-1}a\bar{b} + x_i\left(\bar{a}\bar{b} + ab\right) + x_{i+1}\bar{a}b,\ i \geq 2 \tag{4.15}$$

If we let $\theta = (\bar{a}b)^{-1}(a\bar{b})$, then we have

$$x_i = (\theta^i/\bar{b})x_0,\ i \geq 1. \tag{4.16}$$

Using the normalizing condition, we obtain

$$x_0 = \frac{b-a}{b}. \tag{4.17}$$

Note that for stability, we require that $a < b$.

Hence, for a stable system, we have

$$x_i = (b\bar{b})^{-1}(b-a)\theta^i,\ i \geq 1. \tag{4.18}$$

4.4.2 Transform Approach

We can use the z-transform approach to obtain the same results as follows:

Let

$$X^*(z) = \sum_{i=0}^{\infty} x_i z^i,\ \ |z| \leq 1.$$

We have

$$x_0 = x_0\bar{a} + x_1\bar{a}b \tag{4.19}$$

$$zx_1 = zx_0 a + zx_1\left(\bar{a}\bar{b} + ab\right) + zx_2\bar{a}b \tag{4.20}$$

$$z^i x_i = z^i x_{i-1}a\bar{b} + z^i x_i\left(\bar{a}\bar{b} + ab\right) + z^i x_{i+1}\bar{a}b,\ i \geq 2 \tag{4.21}$$

Taking the sum over i from 0 to ∞ we obtain

$$X^*(z) = X^*(z)[\bar{a}\bar{b} + ab + z^{-1}\bar{a}b + za\bar{b}] + x_0[\bar{a} + za - \bar{a}\bar{b} - ab - z^{-1}\bar{a}b - za\bar{b}]. \quad (4.22)$$

After routine algebraic operations we have

$$X^*(z) = x_0 \frac{u}{v}, \quad (4.23)$$

where

$$\begin{aligned} u &= z(\bar{a} - \bar{a}\bar{b} - ab) - \bar{a}b + z^2(a - a\bar{b}) \\ &= b(za + \bar{a})(z - 1), \end{aligned}$$

and

$$\begin{aligned} v &= z(1 - \bar{a}\bar{b} - ab) - \bar{a}b - z^2 a\bar{b} \\ &= z(\bar{a}b + a\bar{b}) - \bar{a}b - z^2 a\bar{b} \\ &= (\bar{a}b - za\bar{b})(z - 1). \end{aligned}$$

This leads to the results that

$$X^*(z) = b(za + \bar{a})[\bar{a}b - za\bar{b}]^{-1} x_0. \quad (4.24)$$

It is known that $X(z)|_{z \to 1} = 1$. Using this and the above equations we obtain

$$x_0 = \frac{b - a}{b}, \quad (4.25)$$

a result obtained earlier using algebraic approach.

We can write $X^*(z) = f_0 + f_1 z^1 + f_2 z^2 + f_3 z^3 + \cdots$. Based on the algebra of z transform we know that $x_i = f_i$, so we can obtain the probability distribution of the number in the system by finding the coefficients of z^i in the polynomial.

4.4.3 Matrix-geometric Approach

Using the matrix-geometric approach of Section 2.9.7.1, we have

$$A_0 = a\bar{b},\ A_1 = \bar{a}\bar{b} + ab \text{ and } A_2 = \bar{a}b,\ A_v = 0,\ v \geq 3.$$

The matrix R is a scalar in this case, we therefore write it as r and it is a solution to the quadratic equation

$$r = a\bar{b} + r(\bar{a}\bar{b} + ab) + r^2 \bar{a}b \quad (4.26)$$

The solutions to this equation are

$$r = 1 \text{ and } r = \theta$$

The r which we want is the smaller of the two which is $r = \theta$ since $\theta < 1$ for stability. It then follows that

$$x_{i+1} = x_i r = x_i \theta \tag{4.27}$$

We can normalize the vector $\boldsymbol{x}$ by using the matrix $B[R]$. After normalization, we obtain the same results as above , i.e.

$$x_0 = \frac{b-a}{b}.$$

Note that here our $\rho < 1$, if the system is stable. This states

$$\frac{a\bar{b}}{b\bar{a}} < 1 \quad \rightarrow \quad \frac{a}{b} < 1.$$

Example:

Consider an example in which packets arrive according to the Bernoulli process with probability $a = 0.2$ that one arrives in a discrete time slot. Let service be geometric with completion probability of $b = 0.6$ in a single time slot for a packet that is receiving service. We assume we are dealing with a single server. In this case we have

$$\theta = r = \frac{0.2 \times 0.4}{0.6 \times 0.8} = \frac{1}{6},$$

so the system is stable.

The probability of finding an empty system $x_0 = \frac{0.6-0.2}{0.6} = \frac{2}{3}$. Based on these calculations we have

$$x_0 = \frac{2}{3}, \; x_k = \frac{1}{0.6}\left(\frac{1}{6}\right)^k, \; k \geq 1.$$

4.4.4 Performance Measures

One of the most important measures for queueing systems is the number of packets waiting in the system for service, i.e. those in the queues plus the ones in service. The distribution of this measure was presented in the previous section. In this section we present other measures that are very commonly used such as mean values, waiting times and delays and duration of busy period.

4.4.4.1 Mean number in system:

The mean number in the system $E[X] = \sum_{i=1}^{\infty} i x_i$ is obtained as

$$E[X] = (b\bar{b})^{-1}(b-a)\left[\theta + 2\theta^2 + 3\theta^3 + \ldots\right]$$

$$= (b\bar{b})^{-1}(b-a)\theta\left[1+2\theta+3\theta^2+\ldots\right]$$

$$= (b\bar{b})^{-1}(b-a)\theta\frac{d(1-\theta)^{-1}}{d\theta}$$

This results in

$$E[X] = \frac{a\bar{a}}{b-a} \tag{4.28}$$

This same result can be obtained by

$$E[X] = \frac{dX(z)}{dz}|_{z\to 1},$$

where $X(z)$ is the $z-$transform obtained in Section 4.4.2.

Continuing with the example of the previous section, with $a = .2$ and $b = .6$, we have $E[X] = \frac{.2\times.8}{.6-.2} = 0.4$.

4.4.4.2 Mean number in queue:

Let the number in the queue be Y, then the mean queue length $E[Y] = \sum_{i=2}^{\infty}(i-1)x_i$ is obtained as

$$E[Y] = E[X]\frac{a}{b} \tag{4.29}$$

Again using the example of the previous section, with $a = .2$ and $b = .6$, we have $E[Y] = .4\times\frac{.2}{.6} = 0.1333$.

4.4.4.3 Waiting time in the queue:

We assume throughout this chapter that we are dealing with FCFS system, unless we say differently. Let W_q be the waiting time in the queue for a customer and let $w_i^q = \Pr\{W_q = i\}$, with $W_q^*(z) = \sum_{i=0}^{\infty} z^i w_i^q$, then

$$w_0^q = x_0 = \frac{b-a}{b} \tag{4.30}$$

and

$$w_i^q = \sum_{v=1}^{i} x_v \binom{i-1}{v-1} b^v (1-b)^{i-v},\ i \geq 1 \tag{4.31}$$

The arguments leading to this equation are as follows: if we have v items in the system, an arriving item will wait i units of time if in the first $i-1$ time units exactly $v-1$ services are completed and the service completion of the vth item (last one ahead) occurs at time i. This is summed over v from 1 to i.

We obtain the z-transform of W_q as follows. If we replace z in $X^*(z)$ with $B^*(z) = bz(1-(1-b)z)^{-1}$, which is the $z-$transform of the geometric service time distribution of each packet, we obtain

$$X^*(B^*(z)) = g_0 + g_1 z + g_2 z^2 + g_3 z^3 + \cdots, \tag{4.32}$$

implying that g_i, the coefficients of z^i in $X(B(z))$ is w_i^q. Hence we have

$$W_q^*(z) = X^*(B^*(z)) = x_0 \frac{bz + \bar{a}(1-z)}{\bar{a}(1-z) + z(b-a)}. \tag{4.33}$$

The mean time in the queue $E[W_q]$ is given as

$$E[W_q] = \frac{a\bar{a}}{b(b-a)} \tag{4.34}$$

Using the previous example once more we have

$$w_0^q = \frac{2}{3},\ w_1^q = \frac{1}{6},\ w_2^q = \frac{1}{12},\ \text{and so on.}$$

The mean value of waiting time in the queue $E[W_q] = \frac{2}{3}$.

We can also use matrix-analytic approach to study W_q. Consider an absorbing DTMC $\{V_n,\ n \geq 0\}$ with state space $\{0, 1, 2, \cdots\}$, where each number in the state space represent the number of items in the queue ahead of an arriving item. We let this DTMC represent a situation in which an arbitrary item only sees items ahead of it until it gets served. Since we are dealing with a FCFS system the number ahead of this item can not grow in number. Let the transition matrix of this DTMC be $P(w_q)$ written as

$$P(w_q) = \begin{bmatrix} 1 & & & & \\ b & \bar{b} & & & \\ & b & \bar{b} & & \\ & & b & \bar{b} & \\ & & & \ddots & \ddots \end{bmatrix}. \tag{4.35}$$

Define a vector $\mathbf{y}^{(0)} = \boldsymbol{x}$ and consider the recursion

$$\mathbf{y}^{(i+1)} = \mathbf{y}^{(i)} P(w_q),\ i \geq 0.$$

Then we have

$$Pr\{W_q \leq i\} = y_0^{(i)},\ i \geq 0.$$

4.4.4.4 Waiting time in system:

Let W be the waiting time in the system with $w_i = Pr\{W = i\}$, and let the corresponding z-transform be $W^*(z) = \sum_{j=1}^{\infty} z^j w_j$. By simple probability it is clear that W is the convolution sum of W_q and service times, hence we have

$$W^*(z) = W_q^*(z)B^*(z) = x_0\frac{(bz+\bar{a}(1-z))bz}{(\bar{a}(1-z)+z(b-a))(1-(1-b)z)}. \tag{4.36}$$

and the mean waiting time in the system $E[W]$ is given as

$$E[W] = E[W_q] + \frac{1}{b}$$

Using the same numerical example with $a = .2$ and $b = .6$ we have $E[W] = \frac{2}{3} + \frac{1}{.6} =$ 2.3333.

The distribution of the waiting time in the system can be obtained very easily by straightforward probabilistic arguments as follows: An item's waiting time in the system will be i if it has waited in the queue for $j < i$ units of time and its service time is $i-j$, i.e. the first $i-j-1$ time units were without service completion and service completion occurred at the $(i-j)^{th}$ time unit. By summing this over j we obtain the waiting time in the system distribution, i.e.

$$w_i = \sum_{j=0}^{i-1} w_j^q (1-b)^{i-j-1} b, \quad i \geq 1. \tag{4.37}$$

4.4.4.5 Duration of a Busy Period:

The busy period is initiated when a customer arrives to an empty queue and it lasts up to the time the system becomes empty for the first time. Let us define B as the duration of a busy period. This can be studied in at least two ways; one using direct probabilistic arguments and the other is matrix-analytic approach using an absorbing Markov chain.

- **Probabilistic arguments:** Let $S(k)$ be the service time of k items and define $s_j^k = Pr\{S(k) = j\}$ and $A(n)$ the number of arrivals in any interval $[m, m+n]$, and define $a_j^v = Pr\{A(j) = v\}$ and $c_i = Pr\{B = i\}$, we then have:

$$a_j^v = \binom{j}{v} a^v (1-a)^{j-v}, \; 0 \leq j \leq v \geq 1. \tag{4.38}$$

$$s_j^1 = b(1-b)^{j-1}, \; s_j^j = b^j, \tag{4.39}$$

$$s_i^j = bs_{i-1}^{j-1} + (1-b)s_{i-1}^j, \; 1 \leq j \leq i \geq 1. \tag{4.40}$$

The busy period can last i units of time if any of the followings occur: the first arrival requires i units of service and no arrivals occur during this period, and this probability is $s_i^1 a_i^0$, or the first arrival requires $i-k$ service time, however during its service some more items, say v arrive and alls these v items require a total of k time units for service and no arrivals occur during the k time units. This now leads to the result:

$$c_i = s_i^1 a_i^0 + \sum_{k=1}^{i-1} s_{i-k}^1 \sum_{v=1}^{k} a_{i-k}^v s_k^v a_k^0, \; i \geq 1. \tag{4.41}$$

We present another probabilistic approach due to Klimko and Neuts [64] and Heimann and Neuts [54] as follows. A busy period, in the case of single arrivals, is initiated by a single item. By definition c_i is the probability that a busy period initiated by an item lasts for i units of time. Let us assume that the first item's service lasts k units of time with probability s_k^1. Let the number of arrivals during this first item's service be v with probability a_k^v. These v items may be considered as initial items of v independent busy periods. Keeping in mind that busy periods in this case is an independent identically distributed random variable, we know that the busy period generated by v items is a v convolution sum of the independent busy periods. Let $c_j^{(v)}$ be the probability that v independent busy periods last j units of time, then we have

$$c_i = \sum_{k=1}^{i} s_k^1 \sum_{v=0}^{k} a_k^v c_{i-k}^{(v)}, \; i \geq 1. \tag{4.42}$$

If we define the z-transform of the busy period as $c^*(z) = \sum_{i=1}^{\infty} z^i c_i, \; |z| \leq 1$, we have

$$\begin{aligned} c^*(z) &= \sum_{k=1}^{\infty} z^k s_k^1 (1 - a + ac^*(z))^k \\ &= \sum_{k=1}^{\infty} (1-b)^{k-1} bz^k (1 - a + ac^*(z))^k \\ &= bg^*(z)[1 + (1-b)g^*(z) + (1-b)^2 g^*(z)^2 + z^3 g^*(z)^3 + \cdots], \end{aligned} \tag{4.43}$$

where $g^*(z) = z(1 - a + ac^*(z))$.
After routine algebraic operations we obtain

$$c^*(z) = \frac{bg^*(z)}{1 - (1-b)g^*(z)}, \; |z| \leq 1. \tag{4.44}$$

Whereas we have presented a more analytic result due to Klimko and Neuts [64] and Heimann and Neuts [54], the first method is easier for computational purposes. For further discussions on these results the reader is referred to Hunter [57]

- **Matrix-analytic approach:** The idea behind this approach is that we consider the DTMC for the Geo/Geo/1 system but modify it at the boundary such that when the system goes into the 0 state it gets absorbed. The DTMC is initiated from state 1. The transition matrix P^* given as:

$$P^* = \begin{bmatrix} 1 & & & & \\ \bar{a}b & \bar{a}\bar{b}+ab & a\bar{b} & & \\ & \bar{a}b & \bar{a}\bar{b}+ab & a\bar{b} & \\ & & \ddots & \ddots & \ddots \end{bmatrix} \tag{4.45}$$

Let the vector $\mathbf{y}^{(n)} = \left[y_0^{(n)}, y_1^{(n)}, \ldots\right]$ be the state vector of the system at time n, with $\mathbf{y}^{(0)} = [0,\ 1,\ 0,\ 0,\ \cdots]$. Now consider the following recursion

$$\mathbf{y}^{(n+1)} = \mathbf{y}^{(n)} P^*. \tag{4.46}$$

The element $y_0^{(n)}$ is the probability that the busy period last for n or less time units, i.e.

$$y_0^{(n)} = Pr\{B \le n\}. \tag{4.47}$$

4.4.4.6 Number Served During a Busy Period:

Another very important measure is the number of packets served during a busy period. Here we adapt the approach used by Kleinrock [63] for the continuous time analogue to our discrete time case.

Let N_b be the number of packets served during a busy period, with $n_b^k = Pr\{N_b = k\},\ k \ge 1$. Further let $N_b^*(z) = \sum_{k=1}^{\infty} z^k n_b^k,\ |z| \le 1$. Since the arrival process is Bernoulli with parameter a and service is geometric with parameter b, we know that if the duration of service of a packet is j then the probability that i packets arrive during that duration will be given by $\binom{j}{i} a^i (1-a)^{j-i}$. Let u_i be the probability that i packets arrive during the service of one packet then we have $u_i = \sum_{j=max(i,1)}^{\infty} \binom{j}{i} a^i (1-a)^{j-i} (1-b)^{j-1} b$. Letting $U(^*z) = \sum_{i=0}^{\infty} z^i u_i$ we have

$$U^*(z) = [1-(1-b)(1-a+az)]^{-1}(1-a+az)b,$$

which can be written as

$$U^*(z) = ba^*(z)[1-(1-b)a^*(z)]^{-1}, \tag{4.48}$$

where $a^*(z) = 1-a+az$, is the z-transform of the Bernoulli arrival process.

A busy period in this Geo/Geo/1 system is initiated by one packet that arrives after an idle period. Then the number of packets that arrive during this first packet's service each generate sub-busy periods of their own, and since these sub-busy periods are independent and identically distributed, their convolution sum makes up the busy period. Hence if the number of packets that arrive during the service of the first packet is defined as A_b and letting $A_b = j$ and further letting $V_1,\ V_2,\ \cdots,\ V_j$ be the number of packets that arrive during the service of each of the packets $1, 2, \cdots, j$, then the number served during the busy period is $1 + V_1 + V_2 + \cdots + V_j$. Hence we have

$$N_b^*(z) = z \sum_{j=0}^{\infty} Pr\{A_b = j\}(N_b^*(z))^j = zU^*(N_b^*(z)),$$

which results in

$$N_b^*(z) = zba^*(N_b^*(z))[1-(1-b)a^*(N_b^*(z)]^{-1}.$$

4.4.4.7 Age Process

The age process has to do with how long the leading packet in the system has been in the system. Let Y_n be how long the leading packet, which is currently receiving service, has been in the system. It is immediately clear that $Y_n = 0$ implies there are no packets in the system. Consider the stochastic process $\{Y_n,\ n \geq 0\}$, where $Y_n = 0, 1, 2, \cdots$. If the arrival and service processes of packets are according to the Geo/Geo/1 definition, then Y_n is a DTMC with the transition matrix P given and

$$P = \begin{bmatrix} \bar{a} & a & & & \\ \bar{a}b & ba & \bar{b} & & \\ (\bar{a})^2 b & ba\bar{a} & ba & \bar{b} & \\ (\bar{a})^3 b & ba(\bar{a})^2 & ba\bar{a} & ba & \bar{b} \\ \vdots & \vdots & \vdots & \vdots & \vdots & \ddots \end{bmatrix}. \tag{4.49}$$

This is a GI/M/1 type DTMC and the system can be analyzed as such.

4.4.4.8 Workload:

Workload is the amount of work that has accumulated for a server to carry out at any time. Let V_n be the workload at time n. The process $\{V_n,\ n \geq 0\}$, $V_n = 0, 1, 2, \cdots$ is a DTMC, and its transition matrix P can be written as

$$P = \begin{bmatrix} \bar{a} & ab & ab\bar{b} & ab(\bar{b})^2 & ab(\bar{b})^3 & \cdots \\ \bar{a} & ab & ab\bar{b} & ab(\bar{b})^2 & ab(\bar{b})^3 & \cdots \\ & \bar{a} & ab & ab\bar{b} & ab(\bar{b})^2 & \cdots \\ & & \bar{a} & ab & ab\bar{b} & \cdots \\ & & & \ddots & \ddots & \ddots \end{bmatrix}. \tag{4.50}$$

This is an M/G/1 type of DTMC, with a special structure. Standard methods can be used to analyze this.

4.4.5 Discrete time equivalent of Little's Law:

Little's law is a very popular result in queueing theory. In continuous time it is the well known result that $L = \lambda W$, where L is the average number in the system, λ is the arrival rate and W is the mean waiting time in the system. The variables L, λ, and W are used in this previous sentence only to present the well known result. They will not be used with the same meaning after this.

Let W be the waiting time in the system as experienced by a packet and X be the number in the system as left behind by a departing packet. Let $w_k = Pr\{W = k\}$, $k \geq 0$, and $x_j = Pr\{X = j\}$, $j \geq 0$. We let the arrival be geometric with parameter a, as before. Let A_k be the number of arrivals during a time interval of length $k \geq 1$. It is clear that

$$Pr\{A_k = v\} = \binom{k}{v} a^v (1-a)^{k-v}, \; 0 \leq v \leq k.$$

Define

$$X^*(z) = \sum_{j=0}^{\infty} z^j x_j, \quad w^*(z) = \sum_{j=1}^{\infty} z^k w_k, \; |z| \leq 1.$$

Then we have

$$\begin{aligned} X^*(z) &= \sum_{k=1}^{\infty} \sum_{v=0}^{k} \binom{k}{v} a^v (1-a)^{k-v} w_k z^v \\ &= \sum_{k=1}^{\infty} w_k \sum_{v=0}^{k} \binom{k}{v} (az)^v (1-a)^{k-v} \\ &= \sum_{k=1}^{\infty} (az + 1 - a)^k w_k. \end{aligned} \tag{4.51}$$

After routine algebraic operations we obtain

$$X^*(z) = w^*(az + 1 - a). \tag{4.52}$$

By taking the differentials of both sides and setting $z \to 1$ we have

$$E[X] = aE[W], \tag{4.53}$$

which is equivalent to $L = \lambda W$ in the case of Little's law in continuous time. This is actually very straightforward for the Geo/Geo/1. We find that this same idea is extendable to the Geo/G/1 system also.

4.4.6 Geo/Geo/1/K Queues

The Geo/Geo/1/K system is the same as the Geo/Geo/1 system studied earlier, except that now we have a finite buffer $K < \infty$. Hence when a packet arrives to find the buffer full, i.e. K packets in the buffer (implying $K+1$ packets in the system including the one currently receiving service), the arriving packet is dropped. There are cases also where in such a situation an arriving packet is accepted and maybe the packet at the head of the queue is dropped, i.e. LCFS discipline. We will consider that later, even though the two policies do not affect the number in the system, just the waiting time distribution is affected. We can view this system as a special case of the BD process or of the Geo/Geo/1 system. We consider the case when $a_i = a,\ 0 \le i \le K,\ a_i = 0,\ \forall i \ge K+1$ and $b_i = b,\ \forall i \ge 1$. In this case, the transition matrix P becomes

$$P = \begin{bmatrix} \bar{a} & a & & & \\ \bar{a}b & \bar{a}\bar{b}+ab & a\bar{b} & & \\ & \bar{a}b & \bar{a}\bar{b}+ab & a\bar{b} & \\ & & \ddots & \ddots & \ddots \\ & & & \bar{a}b & \bar{a}\bar{b}+a \end{bmatrix}. \tag{4.54}$$

This is a finite space DTMC and methods for obtaining its stationary distribution have been presented in Section 2.9.1. Even though we can use several methods to analyze this Markov chain, we present only the algebraic approach, which is effective here by nature of its structure – tridiagonal structure with repeating rows– and briefly discuss the z-transform approach.

$$x_0 = x_0\bar{a} + x_1\bar{a}b \tag{4.55}$$

$$x_1 = x_0 a + x_1\left(\bar{a}\bar{b} + ab\right) + x_2\bar{a}b \tag{4.56}$$

$$x_i = x_{i-1}a\bar{b} + x_i\left(\bar{a}\bar{b} + ab\right) + x_{i+1}\bar{a}b,\ 2 \le i \le K \tag{4.57}$$

$$x_{K+1} = x_k a\bar{b} + x_{K+1}\left(\bar{a}\bar{b} + a\right) \tag{4.58}$$

If we let $\theta = \frac{a\bar{b}}{\bar{a}b}$, then we have

$$x_i = \left(\frac{\theta^i}{\bar{b}}\right)x_0,\ 1 \le i \le K \tag{4.59}$$

$$x_{K+1} = \frac{a\bar{b}}{\bar{a}b}x_K = \theta^K \frac{a}{\bar{a}b}x_0 \tag{4.60}$$

After normalization, we obtain

$$x_0 = \left[1 + \sum_{i=1}^{K}\frac{\theta^i}{\bar{b}} + \frac{a\theta^K}{\bar{a}b}\right]^{-1} \tag{4.61}$$

Because we have two boundaries in this DTMC at 0 and $K+1$, if we use the z-transform approach we will end up with $X^*(z)$ in terms of the parameters a, b, x_0 and x_{K+1}. We present the z-transform and leave it to the reader to find out how to obtain the stationary distributions from that.

Busy Period

Using the same idea as in the case of the Geo/Geo/1 case but with a slight modification we show that the duration of the busy period for the Geo/Geo/1/K case is a PH type distribution. We begin our discussion as follows: consider an absorbing Markov chain with $K+1$ transient states $\{1,2,\cdots,K+1\}$ and one absorbing state $\{0\}$. Let T be the substochastic matrix representing the transient states transitions and given by

$$T = \begin{bmatrix} \bar{a}\bar{b}+ab & a\bar{b} & & \\ \bar{a}b & \bar{a}\bar{b}+ab & a\bar{b} & \\ & \ddots & \ddots & \ddots \\ & & \bar{a}b & \bar{a}\bar{b}+a \end{bmatrix}. \quad (4.62)$$

Let $\mathbf{t} = \mathbf{1} - T\mathbf{1}$ and also let $\mathbf{y} = [1,0,\ldots]$ then b_i the probability that the busy period duration is i is given as

$$b_i = \mathbf{y}T^{i-1}\mathbf{t},\ i \geq 1. \quad (4.63)$$

The argument here is that the busy period is initiated by the DTMC starting from state $\{1\}$, i.e. due t an arrival and the busy period ends when the system goes into state $\{0\}$, the absorbing state.

The mean duration of the busy period $E[B]$ is simply given as

$$\mu_B = \mathbf{y}(I-T)^{-1}\mathbf{t}. \quad (4.64)$$

We can also study the distribution of the number of packets served during a busy period as a phase type distribution.

Waiting Time in the Queue

Distribution of the waiting time in the queue can also be studied as a PH distribution. Since we are dealing with a FCFS system of service only the items that were ahead of the targeted item get served ahead of it. Because this is a finite buffer system a packet may not receive service at all if it arrives to find the buffer full. The probability of not receiving service is x_{K+1}. Now we focus only on those packets that receive service. Let T be the substochastic matrix representing the transient states transitions, i.e. states $\{1,2,\cdots,K\}$, with state $\{0\}$ as the absorbing state. Then we have

$$T = \begin{bmatrix} \bar{b} & & & \\ b & \bar{b} & & \\ & \ddots & \ddots & \\ & & b & \bar{b} \end{bmatrix}. \quad (4.65)$$

The arriving packet that gets served arrives to find i packets ahead of it with probability $\frac{x_i}{1-x_{K+1}}$. Let us write $\boldsymbol{\alpha} = (1-x_{K+1})^{-1}[x_1, x_2, \ldots, x_K]$. Let $w_i^q = Pr\{W_q = i\}$ we then have

$$w_0^q = (1-x_{K+1})^{-1} x_0, \tag{4.66}$$

and

$$w_i^q = \boldsymbol{\alpha} T^{i-1}\mathbf{t},\ i \geq 1. \tag{4.67}$$

The mean waiting time in the queue for those packets that get served is given as

$$E[W_q] = \boldsymbol{\alpha}(I-T)^{-2}\mathbf{1}. \tag{4.68}$$

4.4.6.1 Departure Process

The departure process from a Geo/Geo/1/K system can be easily represented using the Markovian arrival process (MAP). Consider the transition matrix P associated with this DTMC. It is clear that at each transition there is either a departure or no departure. The matrix P can be partitioned into two matrices D_0 and D_1 as follows.

$$D_0 = \begin{bmatrix} \bar{a} & a & & & \\ & \bar{a}\bar{b} & a\bar{b} & & \\ & & \bar{a}\bar{b} & a\bar{b} & \\ & & & \ddots & \ddots \\ & & & & \bar{a}\bar{b}+a\bar{b} \end{bmatrix}, \quad D_1 = \begin{bmatrix} 0 & 0 & & & \\ \bar{a}b & ab & & & \\ & \bar{a}b & ab & & \\ & & \ddots & \ddots & \\ & & & \bar{a}b & ab \end{bmatrix}.$$

It is clear that $D = D_0 + D_1 = P$. The elements of matrices $(D_k)_{i,j}$ represent the probability of $k = 0, 1$ departures with transitions from i to j. Hence the departure process in a Geo/Geo/1/K system is a MAP and the results from MAP can be used to study this process.

4.5 Geo/G/1 Queues

So far, most of the queues we have studied in discrete time have resulted in univariate DTMCs and both the interarrival and service times are geometric which has the *lack of memory property (lom)*. One of the challenges faced by queueing theorists in the 1950*s* and early 1960*s* was that most queueing systems observed in real life did not have geometric (or exponential) service times and hence the *lom* property assumption for service times was not valid. However, introducing more general service times which do not have the *lom* are difficult to model for a queueing system because then we need to include a supplementary variable in order to capture the Markovian property. By introducing a supplementary variable we end up with a bivariate DTMC, which in theory should be manageable, given the results already discussed in Chapter 2, and which were already available in those days. However, because a general distribution does not necessarily have a finite support then the range of the supplementary variable will be unbounded. This led to a different problem. Until Kendall [62] introduced the method of imbedded Markov chains before this problem became practically feasible to analyze and obtain numerical results. In what follows we present both the method of supplementary variables and the imbedded Markov chain approach for the Geo/G/1 system. The imbedded Markov chain observes the system at points of events, e.g. points of arrivals or points of departures. This way we are able to obtain a Markov chain with usually only a single index which has infinite range. With this approach even though information in between the events is not known directly they can be computed or recovered using renewal theory. The Geo/G/1 queue is studied as an imbedded Markov chain.

Consider a single server queue with geometric interarrival times with probability of an arrival at any point in time given as a and $\bar{a} = 1 - a$. This system has a general service time S with pmf of s_j =Pr$\{S = j,\ j \geq 1\}$, having a mean of μ^{-1}. We define the following z−transforms

$$A^*(z) = 1 - a + za, \text{ and } S^*(z) = \sum_{j=1}^{\infty} z^j s_j.$$

4.5.1 Supplementary Variable Technique

At time n let X_n be the number of packets in the system and J_n the remaining service time of the packet in service. It is immediately clear that $\{(X_n, J_n), n \geq 0\}$ is a bivariate DTMC, keeping in mind that J_n has no value when $X_n = 0$. Its transition matrix is much more difficult to visualize because then both variables are unbounded. Hence we are interested in $x^n_{i,j} = Pr\{X_n = i, J_n = j\}$, letting $x^n_0 = Pr\{X_n = 0\}$. First we write out the related Chapman-Kolmogorov equations as follows:

$$x_0^{n+1} = \bar{a}(x_0^n + x_{1,1}^n) \tag{4.69}$$

$$x_{1,j}^{n+1} = as_j x_0^n + as_j x_{1,1}^n + \bar{a}x_{1,j+1}^n + \bar{a}s_j x_{2,1}^n,\ j \geq 1 \tag{4.70}$$

$$x_{i,j}^{n+1} = as_j x_{i-1,1}^n + as_j x_{i,1}^n + \bar{a}x_{i-1,j+1}^n + \bar{a}s_j x_{i+1,1}^n,\ i \geq 1,\ j \geq 1. \tag{4.71}$$

Next we present two approaches for handling the set of equations in order to obtain the stationary distribution of the system. The two approaches are z-transform and matrix-analytic approaches.

4.5.1.1 Transform Approach

The associated transition matrix can be written as follows: Let the z-transform of the process (X_n, J_n) be

$$X_n^*(z_1,z_2) = x_0 + \sum_{i=1}^{\infty}\sum_{j=1}^{\infty} z_1^i z_2^j x_{i,j}^n, \quad |z_1| \leq 1, \ |z_2| \leq 1.$$

Minh [78] and Dafermos and Neuts [35] have derived the transforms of this system under time-inhomogeneous conditions. Here we focus on stationary distributions only. In that effect we assume the system is stable and consider the case where

$$x_{i,j} = x_{i,j}^n|_{n\to\infty}, \quad x_0 = x_0^n|_{n\to\infty},$$

hence we have

$$X^*(z_1,z_2) = X_n^*(z_1,z_2)|_{n\to\infty}.$$

For this stationary system it was shown by Chaudry [32] that

$$X^*(z_1,z_2) = \frac{z_2 a z_1(1-z_1)x_0\{S(z_1) - S^*(1-a+az_1)\}}{[S^*(1-a-az_1)-z_1]\{z_2-(1-a+az_1)\}}. \tag{4.72}$$

By setting $z_2 = 1$ in $X^*(z_1,z_2)$ we can get the $z-$transform of the marginal distribution, i.e. $y_i = \sum_{j=1}^{\infty} x_{i,j}, \ j \geq 1$. Knowing that $x_0 = 1-\rho, \ \rho = a\mu$, setting $z_2 = 1$ and replacing $z = z_1$ we have $Y(z) = \sum_{i=0}^{\infty} z^i y_i$ which is given as

$$Y^*(z) = \frac{(1-\rho)(1-z)S^*(1-a+za)}{S^*(1-a+za)-z}, \tag{4.73}$$

a result that will be shown through the use of supplementary variables later.

4.5.1.2 Matrix-analytic approach

In order to take full advantage of MAM we need to write the service times distribution as a PH distribution, more of an IPH distribution, based on the elapsed time approach. Let the IPH representation be $(\boldsymbol{\beta}, S)$, with $\mathbf{s} = \mathbf{1} - S\mathbf{1}$, where

$$\boldsymbol{\beta} = [1, \ 0, \ 0, \ \cdots], \ \text{and} \ S_{i,j} = \begin{cases} \tilde{s}_i & j = i+1, \\ 0 & \text{otherwise}, \end{cases}$$

where $\tilde{s}_i = \frac{u_i}{u_{i-1}}, \ u_i = 1 - \sum_{v=1}^{i} s_v, \ i \geq 1, \ u_0 = 1$. Based on the Chapman-Kolmogorov equations presented above we can write the transition matrix of the DTMC as

$$P = \begin{bmatrix} B & C & & & \\ E & A_1 & A_0 & & \\ & A_2 & A_1 & A_0 & \\ & & A_2 & A_1 & A_0 \\ & & & \ddots & \ddots & \ddots \end{bmatrix},$$

where

$$B = \bar{a},\ C = a\boldsymbol{\beta} = [a,\ 0,\ 0,\ \cdots],\ E = \bar{a}\mathbf{s} = \bar{a}[1-\tilde{s}_1,\ 1-\tilde{s}_2,\ 1-\tilde{s}_3,\ \cdots]^T$$

$$A_0 = aS = \begin{bmatrix} 0 & a\tilde{s}_1 & 0 & \cdots \\ 0 & 0 & a\tilde{s}_2 & \cdots \\ & 0 & 0 & \ddots \\ & & \ddots & \end{bmatrix},$$

$$A_1 = a\mathbf{s}\boldsymbol{\beta} + \bar{a}S = \begin{bmatrix} a(1-\tilde{s}_1) & \bar{a}\tilde{s}_1 & 0 & 0 & \cdots \\ a(1-\tilde{s}_2) & 0 & \bar{a}\tilde{s}_2 & 0 & \cdots \\ a(1-\tilde{s}_3) & 0 & 0 & \bar{a}\tilde{s}_3 & \cdots \\ \vdots & \vdots & \vdots & \vdots & \ddots \end{bmatrix},$$

$$A_2 = \bar{a}\mathbf{s}\boldsymbol{\beta} = \begin{bmatrix} \bar{a}(1-\tilde{s}_1) & 0 & 0 & \cdots \\ \bar{a}(1-\tilde{s}_2) & 0 & 0 & \cdots \\ \bar{a}(1-\tilde{s}_3) & 0 & 0 & \cdots \\ \vdots & \vdots & \vdots & \vdots \end{bmatrix}.$$

The G matrix for a stable version of this DTMC QBD is given by the solution to the matrix quadratic equation

$$G = A_2 + A_1G + A_0G^2, \tag{4.74}$$

and this solution is of the form

$$G = \begin{bmatrix} 1 & 0 & 0 & 0 & \cdots \\ 1 & 0 & 0 & 0 & \cdots \\ 1 & 0 & 0 & 0 & \cdots \\ \vdots & \vdots & \vdots & \cdots & \vdots \end{bmatrix}. \tag{4.75}$$

By virtue of the fact that the matrix A_2 is rank one it follows that the corresponding R matrix is given as

$$R = aS[I - a\mathbf{s}\boldsymbol{\beta} - \bar{a}S - aS\mathbf{1}\mathbf{e}_1^T]^{-1}. \tag{4.76}$$

Whereas this result looks elegant and can give us the joint distribution of the number of packets in the system and the remaining interarrival times, we have to truncate R at some point in order to carry out the computations. However, if the general service time has a finite support, i.e. $s_j = 0,\ j < n_s < \infty$, then we can just apply the matrix-geometric method directly because all the block matrices B, C, E and $A_k,\ k = 0, 1, 2$ will all be finite. This situation will be presented later when dealing with the GI/G/1 system which is a generalization of the Geo/G/1 system.

If however, for the case where the service distribution support is not finite, and information about the number in the system as seen by an arriving packet is all we need then we use the imbedded Markov chain approach.

4.5.2 Imbedded Markov Chain Approach

Let X_n be the number of customers left behind by the departing nth customer, and A_n be the number of arrivals during the service of the nth customer. Then we have

$$X_{n+1} = (X_n - 1)^+ + A_{n+1}, \tag{4.77}$$

where $(U-V)^+ = \max\{0, U-V\}$.

This model is classified as *the late arrival with delayed access system* by Hunter [57].

Let $a_v = \Pr\{A_n = v, v \geq 0\}$, $\forall n$. Then a_v is given as

$$a_v = \sum_{k=v}^{\infty} s_k \binom{k}{v} a^v (1-a)^{k-v}, \ v \geq 0. \tag{4.78}$$

Let P be the transition matrix describing this Markov chain, then it is given as

$$P = \begin{bmatrix} a_0 & a_1 & a_2 & \cdots \\ a_0 & a_1 & a_2 & \cdots \\ & a_0 & a_1 & \cdots \\ & & a_0 & \cdots \\ & & & \ddots \end{bmatrix} \tag{4.79}$$

For steady state to exist, we need that $\rho = \sum_{j=1}^{\infty} j a_j < 1$. If we assume that steady state exists, then we have the invariant vector $\boldsymbol{x} = [x_0, x_1, \ldots]$ as

$$\boldsymbol{x} = \boldsymbol{x}P, \ \boldsymbol{x}\mathbf{1} = 1.$$

This can be written as

$$x_i = x_0 a_i + \sum_{j=1}^{i+1} x_j a_{i-j+1}, \ i \geq 0. \tag{4.80}$$

Now define the probability generating functions

$$X^*(z) = \sum_{i=0}^{\infty} x_i z^i, \text{ and } A^*(z) = \sum_{i=0}^{\infty} a_i z^i = S^*(1-a+za), \ |z| \leq 1,$$

we obtain

$$X^*(z) = \frac{x_0(1-z)A^*(z)}{A^*(z) - z}. \tag{4.81}$$

Using the fact that $X^*(1) = 1$, $A^*(1) = 1$ and $(A^*)'(1) = a/\mu$, we find that

$$x_0 = 1 - \rho \tag{4.82}$$

and

$$X^*(z) = \frac{(1-\rho)(1-z)A^*(z)}{A^*(z)-z} = \frac{(1-\rho)(1-z)S^*(1-a+za)}{S^*(1-a+za)-z}. \quad (4.83)$$

The mean queue length $E[X]$ is given as

$$E[X] = \rho + \frac{a^2}{2(1-\rho)}\sigma_s^2, \quad (4.84)$$

where σ_s^2 is the variance of the service time and $\rho = a/\mu$.

4.5.2.1 Distribution of number in the system

The probability of the number of packets left behind in the system by a departing packet x_i, $i = 0,1,2,\cdots$, can be obtained by finding the coefficients of z^i in the polynomial associated with $X(z)$, but this could be onerous. The simplest approach is to apply the recursion by Ramaswami [89] as follows. If the system is stable, then the G matrix (a scalar in this case) associated with the M/G/1-type system is a scalar and has the value of 1. Since x_0 is known to be $1-\rho$, then we can successfully apply the recursion to obtain this probability distribution.

Let $\tilde{a}_v = \sum_{i=v}^{\infty} a_i$, $v \geq 0$, then

$$x_i = (1-\tilde{a}_1)^{-1}[x_0\tilde{a}_i + (1-\delta_{i,1})\sum_{j=1}^{i-1} x_j\tilde{a}_{i+1-j}], \; i \geq 1. \quad (4.85)$$

The recursion is initiated by $x_0 = 1-\rho$.

4.5.2.2 Waiting Time:

Let W be the waiting time in the system, with $w_k = Pr\{W = k\}$ and $W^*(z) = \sum_{k=0}^{\infty} z^k w_k$. We know that the number of items left behind by a departing item is the number of arrivals during its waiting time in the system, hence

$$x_i = \sum_{k=i}^{\infty} \binom{k}{i} a^i (1-a)^{k-i} w_k, \; i \geq 0. \quad (4.86)$$

By taking the $z-$ transform of both sides we end up with

$$X^*(z) = \sum_{k=0}^{\infty} (1-a+az)^k w_k = W^*(1-a+za). \quad (4.87)$$

Hence

$$W^*(1-a+za) = \frac{(1-\rho)(1-z)S^*(1-a+za)}{S^*(1-a+za)-z}. \tag{4.88}$$

4.5.2.3 Workload:

Let $V_n, n \geq 0$ be the workload in the system at time n, with $V_n = 0, 1, 2, \cdots$. It is straightforward to show that V_n is a DTMC and its transition matrix is an M/G/1 type with the following structure

$$P = \begin{bmatrix} c_0 & c_1 & c_2 & c_3 & \cdots \\ c_0 & c_1 & c_2 & c_3 & \cdots \\ & c_0 & c_1 & c_2 & \cdots \\ & & c_0 & c_1 & \cdots \\ & & & \ddots & \ddots \end{bmatrix}, \tag{4.89}$$

with

$$c_0 = 1-a, \; c_j = ab_j, \; j \geq 1.$$

If the system is stable, i.e. if $1 > \sum_{j=1}^{\infty} jc_j$, then we have a stationary distribution $\mathbf{v} = [v_0, \; v_1, \; v_2, \; \cdots]$ such that

$$\mathbf{v} = \mathbf{v}P, \;\; \mathbf{v1} = 1.$$

This leads to the equation

$$v_i = c_i v_0 + \sum_{j=1}^{i+1} v_j c_{i-j+1}, \;\; i \geq 0. \tag{4.90}$$

It is immediately clear that we can set this up in such a manner that allows us to use Ramaswami's recursion to obtain the stationary distribution at arbitrary times. By letting

$$C^*(z) = \sum_{i=0}^{\infty} z^i c_i, \;\; V^*(z) = \sum_{i=0}^{\infty} z^i v_i, \; |z| \leq 1,$$

we have

$$V^*(z) = \frac{v_0[(z-1)C^*(z)]}{z-C^*(z)}, \; |z| \leq 1. \tag{4.91}$$

Since the resulting equations are exactly of the same form as the ones for the number in the system we simply apply the techniques used earlier for similar structure problems to obtain key measures.

4.5.2.4 Age Process:

Here we study the age process, but first we re-write the service time in the form of elapsed time. Consider the service time distribution $\{b_1, b_2, b_3, \cdots\}$. It can be written as an infinite PH (IPH) distribution represented as (β, S) where

$$\beta = [1,\ 0,\ 0,\ \cdots],\ \text{and}\ S_{i,j} = \begin{cases} \tilde{b}_i, & j = i+1 \\ 0, & \text{otherwise}, \end{cases}$$

where

$$\tilde{b}_i = \frac{e_i}{e_{i-1}},\ e_i = 1 - \sum_{v=1}^{i} b_v,\ e_0 = 1.$$

It was pointed out in Chapter 3 that

$$b_k = \beta S^{k-1}\mathbf{s},\ k \geq 1.$$

At time $n \geq 0$, let Y_n be the age of the leading packet in the system, i.e. the one currently receiving service, and let J_n be its elapsed time of service. The bivariate process $\{(Y_n, J_n), n \geq 0\}$ is a DTMC of the GI/M/1 type with transition matrix P written as

$$P = \begin{bmatrix} C_0 & C_1 & & & & \\ B_1 & A_1 & A_0 & & & \\ B_2 & A_2 & A_1 & A_0 & & \\ B_3 & A_3 & A_2 & A_1 & A_0 & \\ \vdots & \vdots & \vdots & \vdots & \vdots & \ddots \end{bmatrix}, \quad (4.92)$$

where,

$$C_0 = \bar{a},\ C_1 = a\beta,\ B_k = \bar{a}^k\mathbf{s},\ A_0 = S,\ A_k = a\bar{a}^{k-1}(\mathbf{s}\beta),\ k \geq 1.$$

With this set up the standard GI/M/1 results can be applied to the age process to obtain necessary measures. However, because we have used the IPH caution has to be exercised because now we are dealing with infinite block matrices. If the service time distribution has a finite support then it is straightforward to analyze the system. When it is not a finite support one needs to refer to the works of Tweedie [99] to understand the conditions of stability.

4.5.2.5 Busy Period:

Let $b_i(j)$ be the probability that the busy period initiated by j customers lasts i units of time. It was shown by Klimko and Neuts [64] and Heimann and Neuts [54] that

$$b_i(1) = \sum_{l=1}^{i} s_1(1) \sum_{k=0}^{l} d_k^l b_{i-1}(k) \quad (4.93)$$

where a_k^l is the probability that k arrivals occurs in l units of time, and $s_i(k)$ is the probability that the service times of k customers lasts exactly i units of time.

4.5.3 Geo/G/1/K Queues

This is the Geo/G/1 system with finite buffer. Let the buffer size be $K < \infty$. We can use the imbedded Markov chain approach to analyze this system.

Let X_n be the number of customers left behind by the departing nth customer, and A_n be the number of arrivals during the service of the nth customer. Then we have

$$X_{n+1} = ((X_n - 1)^+ + A_{n+1})^- \tag{4.94}$$

where $(U)^-$ =min$\{K+1,\ U\}$ and $(V)^+$ =max$\{0, V\}$.

This model is classified as *the late arrival with delayed access system* by Hunter [57].

Let a_v =Pr$\{A_n = v,\ v \geq 0\}$, $\forall n$. Then a_v is given as

$$a_v = \sum_{k=v}^{\infty} s_k \binom{k}{v} a^v (1-a)^{k-v},\ v \geq 0. \tag{4.95}$$

Let P be the transition matrix describing this Markov chain, then it is given as

$$P = \begin{bmatrix} a_0 & a_1 & a_2 & \cdots & a_{K-1} & \tilde{a}_K \\ a_0 & a_1 & a_2 & \cdots & a_{K-1} & \tilde{a}_K \\ & a_0 & a_1 & \cdots & a_{K-2} & \tilde{a}_{K-1} \\ & & a_0 & \cdots & a_{K-3} & \tilde{a}_{K-2} \\ & & & \ddots & \ddots & \vdots \\ & & & & a_0 & \tilde{a}_1 \end{bmatrix}, \tag{4.96}$$

where $\tilde{a}_j = \sum_{v=j}^{\infty} a_v$.

Letting the stationary distribution be $\boldsymbol{x} = [x_0,\ x_1,\ \cdots,\ x_{K+1}]$, we have

$$x_j = a_j x_0 + \sum_{v=1}^{j+1} a_{j-v+1} x_v,\ 0 \geq j \geq K-1,\ \text{ and} \tag{4.97}$$

$$x_K = \tilde{a}_K x_0 + \sum_{v=1}^{K} a_{K-v+1} x_v. \tag{4.98}$$

$$\tag{4.99}$$

We can apply the standard finite Markov chain analysis methods discussed in Chapter 2 for finding the stationary distribution, i.e. applying

$$\boldsymbol{x} = \boldsymbol{x}P,\ \boldsymbol{x}\mathbf{1} = 1.$$

A very interesting thing to note about this system is that if its traffic intensity is less than 1, then we can actually use the Ramaswami's recursion idea for the M/G/1. This is because the g given as the minimal non-negative solution to the equation $g = \sum_{k=0}^{\infty} g^k a_k$ has a solution that $g = 1$ and by applying the recursion we have

$$x_i = (1-\tilde{a}_1)^{-1}[x_0\tilde{a}_i + (1-\delta_{i,1})\sum_{v=1}^{i-1} x_v\tilde{a}_{i+1-v}], \ 1 \le i \le K. \tag{4.100}$$

However, we need to determine x_0 first but this is done using the normalization equation.

4.6 GI/Geo/1 Queues

The GI/Geo/1 model is essentially the "dual" of the Geo/G/1. Suppose the arrivals are of the general independent type with inter-arrival times G having a pdf of $g_n = \Pr\{G = n, \ n \ge 1\}$ whose mean is λ^{-1} and the service are of the geometric distribution with parameters s and $\bar{s}$ and with mean service time s^{-1}. Let $G^*(z) = \sum_{i=1}^{\infty} z^i g_i, \ |z| \le 1$. Because the inter-arrival time is general and the service is geometric, we study this also using supplementary variables and the imbedded Markov chain which is imbedded at the point of arrival.

4.6.1 Supplementary Variable Technique

Define X_n and K_n as the number of packets in the system and the remaining interarrival time at time $n \ge 0$. Further let $x_{i,j}^{(n)} = Pr\{X_n = i, K_n = j\}$. Then we have

$$x_{0,j}^{(n+1)} = x_{0,j+1}^{(n)} + x_{1,j+1}^{(n)} s, \ \ j \ge 1, \tag{4.101}$$

$$x_{1,j}^{(n+1)} = x_{0,1}^{(n)} g_j + x_{1,1}^{(n)} s g_j + x_{1,j+1}^{(n)} \bar{s} + x_{2,j+1}^{(n)} s, \ \ j \ge 1. \tag{4.102}$$

$$x_{i,j}^{(n+1)} = x_{i-1,1}^{(n)} \bar{s} g_j + x_{i,1}^{(n)} s g_j + x_{i,j+1}^{(n)} \bar{s} + x_{i+1,j+1}^{(n)} s, \ \ j \ge 2. \tag{4.103}$$

4.6.1.1 Matrix-analytic approach

We have a bivariate DTMC with the transition matrix P written as

$$P = \begin{bmatrix} B & C & & & \\ A_2 & A_1 & A_0 & & \\ & A_2 & A_1 & A_0 & \\ & & \ddots & \ddots & \ddots \end{bmatrix},$$

with

$$B = \begin{bmatrix} 0 & 0 & 0 & \cdots \\ 1 & 0 & 0 & \cdots \\ 0 & 1 & 0 & \cdots \\ & & \ddots & \end{bmatrix}, \quad A_1 = \begin{bmatrix} sg_1 & sg_2 & sg_3 & \cdots \\ \bar{s} & 0 & 0 & \cdots \\ 0 & \bar{s} & 0 & \cdots \\ & & \ddots & \end{bmatrix},$$

$$C = \begin{bmatrix} g_1 & g_2 & g_3 & \cdots \\ 0 & 0 & 0 & \cdots \\ 0 & 0 & 0 & \cdots \\ \vdots & \vdots & \vdots & \cdots \end{bmatrix},$$

$$A_2 = sB, \quad A_0 = \bar{s}C.$$

We can now easily apply the matrix-geometric results for the QBD to this system, given all the conditions required by the infinite blocks of P are satisfied. With

$$R = A_0 + RA_1 + R^2 A_2,$$

and the known properties of R we can see that it has the following structure

$$R = \begin{bmatrix} r_1 & r_2 & r_3 & \cdots \\ 0 & 0 & 0 & \cdots \\ 0 & 0 & 0 & \cdots \\ \vdots & \vdots & \vdots & \cdots \end{bmatrix},$$

where

$$r_k = \bar{s}g_k + r_1 s g_k + s r_1 r_{k+1}, \quad k \geq 1. \tag{4.104}$$

Whereas this result looks elegant and can give us the joint distribution of the number of packets in the system and the remaining interarrival times, we have to truncate R at some point in order to carry out the computations. However, if the general interarrival time has a finite support, i.e. $a_j = 0, \ j < K_a = n_t < \infty$, then we can just apply the matrix-geometric method directly because all the block matrices B, C, E and $A_k, \ k = 0, 1, 2$ will all be finite. The case of finite support, i.e. $n_t < \infty$ is given as follows.

If $n_t < \infty$, then we have

$$r_k = \bar{s}g_k + r_1 s g_k + s r_1 r_{k+1}, \quad 1 \leq k < n_t, \tag{4.105}$$

$$r_{n_t} = \bar{s}g_{n_t} + r_1 s g_{n_t}. \tag{4.106}$$

Applying the last equation and working backwards we have

$$r_{n_t-1} = \bar{s}g_{n_t-1} + r_1 s g_{n_t-1} + s r_1 r_{n_t}$$

$$= \bar{s}g_{n_t-1} + r_1 s g_{n_t-1} + s r_1 \bar{s} g_{n_t} + s^2 (r_1)^2 g_{n_t}. \quad (4.107)$$

By repeated application we find that r_1 is a polynomial equation of the form

$$r_1 = a_0 + a_1 r_1 + a_2 (r_1)^2 + a_3 (r_1)^3 + \cdots + a_{n_t} (r_1)^{n_t}, \quad (4.108)$$

where $a_j,\ j = 0, 1, \cdots, n_t$ are known constants. For example when $n_t = 2$ we have

$$a_0 = \bar{s}g_1,\ a_1 = s g_1 + \bar{s} g_2, \text{ and } a_2 = s^2 g_2.$$

We solve this polynomial for the minimum r_1 which satisfies $0 < r_1 < 1$. Once r_1 is known we can then solve for r_{n_t} and then proceed to obtain all the other $r_j,\ j = n_t - 1, n_t - 2, \cdots, 3, 2$. This case will be presented later when dealing with the general case of GI/G/1 system.

If however, for the case where the interarrival time does not have a finite support, and we need information about the number in the system as seen by a departing packet then we use the imbedded Markov chain approach.

4.6.2 Imbedded Markov Chain Approach

Let Y_n be the number of packets in the system at the arrival time of the nth packet, and B_n as the number of packets served during the inter-arrival time of the nth packet. Then we have

$$Y_{n+1} = Y_n + 1 - B_n,\ Y_n \geq 0,\ B_n \leq Y_n + 1 \quad (4.109)$$

Further, let us define b_j as the probability that j packets are served during the $(n+1)$th inter-arrival time if the system is found empty by the $(n+1)$th arriving packet and a_j is the corresponding probability if this packet arrives to find the system not empty with j packets served during its inter-arrival time. Then we have

$$a_j = \sum_{k=j}^{\infty} g_k \binom{k}{j} s^j (1-s)^{k-j},\ j \geq 0 \quad (4.110)$$

$$b_j = 1 - \sum_{i=0}^{j} a_i,\ j \geq 0 \quad (4.111)$$

The transition matrix describing this Markov chain is given by P as follows

$$P = \begin{bmatrix} b_0 & a_0 & & & \\ b_1 & a_1 & a_0 & & \\ b_2 & a_2 & a_1 & a_0 & \\ \vdots & \vdots & \vdots & \ddots & \ddots \end{bmatrix}. \quad (4.112)$$

This system is called the *early arrival system* by Hunter [57].

Let $\boldsymbol{x} = [x_0, x_1, \ldots]$ be the invariant vector associated with this Markov chain. It is simple to use either the matrix-geometric approach or the transform approach to obtain this vector.

4.6.2.1 Matrix-geometric approach:

It is straightforward to see that the matrix P here has the same structure as the GI/M/1-type presented in the previous chapter. The GI/Geo/1 queue is the discrete time analogue of the GI/M/1 system. The descriptor GI/M/1-type got its name from the classical GI/M/1 queue transition probability matrix structure. We note that this GI/Geo/1 queue transition matrix has scalar entries in P and as such the corresponding R is a scalar r.

Provided that the system is stable, i.e. $\rho = \left[\sum_{j=1}^{\infty} jb_j\right]^{-1} < 1$, then there exists r which is the minimal non-negative solution to the polynomial equation

$$r = \sum_{i=0}^{\infty} a_i r^i. \tag{4.113}$$

The resulting solution gives us the geometric form of the solution for $\boldsymbol{x}$ as follows:

$$x_{i+1} = x_i r, \; i \geq 0. \tag{4.114}$$

After solving for the boundary value $x_0 = 1 - \rho = 1 - r$ through $x_0 = \sum_{j=0}^{\infty} x_j b_j$ and applying the normalization equation $\sum_{j=0}^{\infty} x_j = 1$, we obtain

$$x_i = (1-r)r^i, \; i \geq 0. \tag{4.115}$$

The results in (4.113) and (4.115) are actually the scalar-geometric solution analogue of the matrix-geometric solution presented in the previous chapter.

4.6.2.2 Transform approach

Consider the matrix P, we can write the steady state equations for this system as

$$x_0 = \sum_{j=0}^{\infty} x_j b_j = \sum_{j=0}^{\infty} x_j (1 - \sum_{v=0}^{j} a_v), \tag{4.116}$$

$$x_i = \sum_{j=1}^{\infty} x_{i+j-1} a_j, \; i \geq 1. \tag{4.117}$$

Adopting the approach in Grass and Harris [49] we can solve the above set of difference equations by using the operator approach. Consider the operator F and write

$$x_{i+1} = Fx_i, \; i \geq 1,$$

then substituting this into the above set of equations for $x_i, \; i \geq 1$, i.e.

$$x_i - [\sum_{j=1}^{\infty} x_{i+j-1} a_j] = 0,$$

we have

$$F - a_0 - Fa_1 - F^2 a_2 - F^3 a_3 - \cdots = 0. \tag{4.118}$$

This leads to the result that $F = \sum_{j=0}^{\infty} F^j a_j$. The characteristic equation for this difference equation is

$$z = \sum_{j=0}^{\infty} z^j a_j. \tag{4.119}$$

However, a_j is the probability that the number of service completions during an interarrival time is j, so by substituting $sz + 1 - s$ into $\sum_{j=0}^{\infty} z^j a_j$ we get

$$z = G^*(sz + 1 - s). \tag{4.120}$$

It can be shown that a solution z^* that satisfies this equation is the same as r which we obtained earlier using the matrix-geometric approach. Hence

$$r = G^*(sr + 1 - s). \tag{4.121}$$

Once again provided the system is stable we only have one positive value that satisfies this equation.

It can be shown that

$$x_i = Kr^i, \; i \geq 0, \tag{4.122}$$

and when this is normalized we find that $K = 1 - r$.

A simple way to obtain r is by iterating the following

$$r_{k+1} = G^*(r_k s + (1-s)) \text{ with } 0 \leq r_0 < 1 \tag{4.123}$$

until $|r_{k+1} - r_k| < \varepsilon$, where ε is a very small positive value, e.g. 10^{-12}.

Waiting time:

- Waiting time in the queue: Let W_q be the waiting time in the queue for a packet, with $w_j^q = Pr\{W_q = j\}, \; j \geq 0$. Since this is a FCFS system, then an arriving packet only has to wait for the packets ahead of it. Let $W_q^*(z) = \sum_{j=0}^{\infty} z^j w_j^q$. It is clear that the $z-$ transform of the probability x_i is $X^*(z) = \sum_{i=0}^{\infty} z^i x_i = (1-r)(1-rz)^{-1}$. Since this is the transform of the number of packets seen by an arriving packet then to obtain $W_q^*(z)$ we simply replace z in $X^*(z)$ with $sz(1-(1-s)z)^{-1}$ leading to

$$W_q^*(z) = (1-r)(1 - rsz(1-(1-s)z)^{-1})^{-1}.$$

Hence

$$W_q^*(z) = \frac{(1-r)(1-(1-s)z)}{1-(1-s)z-rsz}. \tag{4.124}$$

- Waiting time in the system: The $z-$ transform of the waiting time in the system is simply a convolution sum of W_q and service times. Let $W^*(z)$ be the $z-$ transform of the waiting time in the system then we have

$$W^*(z) = (1-r)sz(1-rsz(1-(1-s)z)^{-1})^{-1}(1-(1-s)z)^{-1}.$$

Hence

$$W^*(z) = \frac{sz(1-r)}{1-(1-s)z-rsz}. \tag{4.125}$$

4.6.3 GI/Geo/1/K Queues

This is the finite buffer case of the GI/Geo/1 system. So let Y_n be the number of packets in the system at the arrival time of the nth packet, and B_n as the number of packets served during the interarrival time of the nth packet. Then we have

$$Y_{n+1} = (Y_n + 1 - B_n)^-, \; Y_n \geq 0, \; B_n \leq Y_n + 1, \tag{4.126}$$

where $(V)^-$ =min(K,V). Further, let us define b_j as the probability that j packets are served during the $(n+1)$th interarrival time if the system is found empty by the $(n+1)$th arriving packet and a_j is the corresponding probability if this packet arrives to find the system not empty with j packets served during its interarrival time. Then we have

$$a_j = \sum_{k=j}^{\infty} g_k \binom{k}{j} s^j (1-s)^{k-j}, \; j \geq 0 \tag{4.127}$$

$$b_j = 1 - \sum_{i=0}^{j} a_i, \; j \geq 0 \tag{4.128}$$

The transition matrix describing this Markov chain is given by P as follows

$$P = \begin{bmatrix} b_0 & a_0 & & & & \\ b_1 & a_1 & a_0 & & & \\ b_2 & a_2 & a_1 & a_0 & & \\ \vdots & \vdots & \vdots & \vdots & \ddots & \\ b_{K-1} & a_{K-1} & a_{K-2} & \cdots & a_1 & a_0 \\ b_K & a_K & a_{K-1} & \cdots & a_2 & a_1 + a_0 \end{bmatrix} \tag{4.129}$$

Let the stationary distribution be $\boldsymbol{x} = [x_0, \; x_1, \; \cdots, \; x_K]$, then we have

$$\boldsymbol{x} = \boldsymbol{x}P, \ \boldsymbol{x}\mathbf{1} = 1.$$

Once more, this can be solved using standard Markov chain techniques presented in Chapter 2.

4.7 Geo/PH/1 Queues

Consider a single server queue with geometric interarrival times with parameter a and PH service with representation $(\boldsymbol{\beta}, S)$ of order n_s. At time n, let X_n and J_n be the number of packets in the system and the phase of service of the packet receiving service. Consider this system as a DTMC with the state space $\{(0) \cup (X_n, J_n), n \geq 0\}, X_n \geq 1, J_n = 1, 2, \cdots, n_s$. This is a QBD with the transition matrix P given as

$$P = \begin{bmatrix} B & C & & & \\ E & A_1 & A_0 & & \\ & A_2 & A_1 & A_0 & \\ & & \ddots & \ddots & \ddots \end{bmatrix},$$

$$B = 1-a, \ E = (1-a)\mathbf{s}, \ C = a\boldsymbol{\beta}, \ A_0 = aS,$$

$$A_1 = a\mathbf{s}\boldsymbol{\beta} + (1-a)S, \ A_2 = (1-a)\mathbf{s}\boldsymbol{\beta}.$$

We can go ahead and use the results for the QBD presented in Chapter 2 to study this system. However, this system has an added feature that we can exploit. The matrix A_2 is rank one, i.e. $(1-a)\mathbf{s}\boldsymbol{\beta}$. Even though we need to compute the matrix R we try and obtain it through matrix G, as follows. Remember that $G = (I-U)^{-1}A_2$, from the last chapter. Hence we have

$$G = (I-U)^{-1}(1-a)\mathbf{s}\boldsymbol{\beta} = \mathbf{d}\boldsymbol{\beta}, \tag{4.130}$$

where $\mathbf{d} = (I-U)^{-1}(1-a)\mathbf{s}$.

Since G is stochastic, if the DTMC is stable, by Theorem 8.5.1 in Latouche and Ramaswami [67], we have

$$G\mathbf{1} = \mathbf{d}\boldsymbol{\beta}\mathbf{1} = \mathbf{1} \ \rightarrow G = \mathbf{1}\boldsymbol{\beta}. \tag{4.131}$$

Using this result and substituting into $R = A_0(I - A_1 - A_0 G)^{-1}$ we obtain

$$R = A_0[I - A_1 - A_0\mathbf{1}\boldsymbol{\beta}]^{-1}, \tag{4.132}$$

an explicit solution for matrix R from an explicit solution for G.

The stationary distribution of this system $x_{i,j} = Pr\{X_n = i, J_n = j\}|_{n\rightarrow\infty}$ and $x_0 = Pr\{X_n = 0\}|_{n\rightarrow\infty}$ leading to $\boldsymbol{x} = [x_0, \ \boldsymbol{x}_1, \ \boldsymbol{x}_2, \ \cdots, \ ]$, where $\boldsymbol{x}_i = [x_{i,1}, \ x_{i,2}, \ \cdots, \ x_{i,n_s}]$, we have

$$\boldsymbol{x} = \boldsymbol{x}P, \ \ \boldsymbol{x}\mathbf{1} = 1,$$

and

$$\boldsymbol{x}_{i+1} = \boldsymbol{x}_i R, \ i \geq 1,$$

with the boundary equations

$$x_0 = x_0(1-a) + \boldsymbol{x}_1 E, \ \boldsymbol{x}_1 = x_0 C + \boldsymbol{x}_1(A_1 + RA_2),$$

and normalization equation

$$x_0 + \boldsymbol{x}_1(I-R)^{-1}\mathbf{1} = 1,$$

we obtain the stationary distribution for this system.

The mean number in the system $E[X]$ is given as

$$E[X] = (\boldsymbol{x}_1 + 2\boldsymbol{x}_2 + 3\boldsymbol{x}_3 + \cdots)\mathbf{1} = \boldsymbol{x}_1(I + 2R + 3R^2 + \cdots)\mathbf{1},$$

which gives

$$E[X] = \boldsymbol{x}_1(I-R)^{-2}\mathbf{1}. \tag{4.133}$$

4.7.1 Waiting Times

For this we assume a FCFS system of service. Let W_q be the waiting time in the queue, with $w_i^q = Pr\{W_q = i\}, \ i \geq 0$ and let $W_q^*(z) = \sum_{j=0}^{\infty} z^j w_j^q, \ |z| \leq 1$, then we have

$$W_q^*(z) = x_0 + \sum_{i=1}^{\infty} \boldsymbol{x}_i (z\boldsymbol{\beta}(I-zS)^{-1}\mathbf{s})^i,$$

which gives

$$W_q^*(z) = x_0 + \boldsymbol{x}_1(z\boldsymbol{\beta}(I-zS)^{-1}\mathbf{s})[I - Rz\boldsymbol{\beta}(I-zS)^{-1}\mathbf{s})]^{-1}. \tag{4.134}$$

Alternatively we may compute w_j^q as follows.

Let $B_k^k = (\mathbf{s}\boldsymbol{\beta})^k, \ k \geq 1, \ B_j^1 = S^{j-1}(\mathbf{s}\boldsymbol{\beta}), \ j \geq 1$, then we have

$$B_j^k = SB_{j-1}^k + (\mathbf{s}\boldsymbol{\beta})B_{j-1}^{k-1}, \ k \geq j \geq 1. \tag{4.135}$$

We then write

$$w_0^q = x_0, \tag{4.136}$$

$$w_j^q = \sum_{v=1}^{j} \boldsymbol{x}_v B_j^v, \ j \geq 1. \tag{4.137}$$

On the other hand we may apply the idea of absorbing DTMC to analyze the waiting time, as follow. Consider a Markov chain $\{(\tilde{X}_n, J_n)\}$ where $\tilde{X}_n$ is the number of packets ahead of a target packet at time n. Since this is a FCFS system, if we let $\tilde{X}_n = 0$ be an absorbing state, then the time to absorption in this DTMC is the waiting time in the queue. The transition matrix for the DTMC is

$$P_w = \begin{bmatrix} 1 & & & \\ \mathbf{s} & S & & \\ & \mathbf{s}\boldsymbol{\beta} & S & \\ & & \mathbf{s}\boldsymbol{\beta} & S \end{bmatrix}.$$

The stationary vector of the number in the system now becomes the intial vector of this DTMC, i.e. let $\mathbf{y}(0) = \boldsymbol{x}$, where $\mathbf{y}(0) = [y_0(0), \mathbf{y}_1(0), \mathbf{y}_2(0), \cdots]$, then we can apply the recursion

$$\mathbf{y}(k+1) = \mathbf{y}(k)P_w, \; k \geq 0.$$

From this we have

$$Pr\{W_q \leq j\} = W_j^q = y_0(j), \; j \geq 1, \tag{4.138}$$

$$W_0^q = y_0(0). \tag{4.139}$$

Waiting time in the system can be obtained directly as a convolution sum of W_q and service time.

4.8 PH/Geo/1 Queues

Consider a single server queue with PH interarrival times with parameter $(\boldsymbol{\alpha}, T)$ of order n_t and geometric service with parameter b representing probability of a service completion in a time epoch when a packet is receiving service. At time n_t, let X_n and J_n be the number of packets in the system and the phase of the interarrival time. Consider this system as a DTMC with the state space $\{(X_n, J_n), n \geq 0\}, X_n \geq 0, J_n = 1, 2, \cdots, n_t$. This is a QBD with the transition matrix P given as

$$P = \begin{bmatrix} B & C & & & \\ E & A_1 & A_0 & & \\ & A_2 & A_1 & A_0 & \\ & & \ddots & \ddots & \ddots \end{bmatrix},$$

$$B = T, \; E = bT, \; C = \mathbf{t}\boldsymbol{\alpha}, \; A_0 = (1-b)(\mathbf{t}\boldsymbol{\alpha}),$$

$$A_1 = b(\mathbf{t}\boldsymbol{\alpha}) + (1-b)T, \; A_2 = bT.$$

We can go ahead and use the results for the QBD presented in Chapter 2 to study this system. However, this system has an added feature that we can exploit. Because the matrix A_0 is rank one, i.e. $(1-b)\mathbf{t}\boldsymbol{\alpha}$, the matrix R can be written explicitly as

$$R = A_0(I - A_1 - \eta A_2)^{-1}, \tag{4.140}$$

where $\eta = sp(R)$, i.e. the spectral radius of matrix R. This is based on Theorem 8.5.2 in Latouche and ramaswami [67]. Even though this is an explicit expression we still need to obtain η. So in a sense, it is not quite an explicit expression. The term η can be obtained by solving for z in the scalar equation

$$z = \boldsymbol{\alpha}(I - A_1 - zA_2)^{-1}\mathbf{1}. \tag{4.141}$$

A detailed discussion of this can be found in [67].

Let $x_{i,j} = Pr\{X_n = i, J_n = j\}|_{n\to\infty}$ be the stationary distribution of this system. Further define $\boldsymbol{x} = [\boldsymbol{x}_0, \boldsymbol{x}_1, \boldsymbol{x}_2, \cdots,]$, where $\boldsymbol{x}_i = [x_{i,1}, x_{i,2}, \cdots, x_{i,n_t}]$, we have

$$\boldsymbol{x} = \boldsymbol{x}P, \ \boldsymbol{x}\mathbf{1} = 1,$$

and

$$\boldsymbol{x}_{i+1} = \boldsymbol{x}_i R, \ i \geq 1,$$

with the boundary equations

$$\boldsymbol{x}_0 = \boldsymbol{x}_0 B + \boldsymbol{x}_1 E, \ \boldsymbol{x}_1 = \boldsymbol{x}_0 C + \boldsymbol{x}_1 (A_1 + RA_2),$$

and normalization equation

$$[\boldsymbol{x}_0 + \boldsymbol{x}_1 (I - R)^{-1}]\mathbf{1} = 1,$$

we obtain the stationary distribution for this system.

The mean number in the system $E[X] = \boldsymbol{x}_1 (I - R)^{-2}\mathbf{1}$.

4.8.1 Waiting Times

To study the waiting time we first need to obtain the distribution of the number in the system as seen by an arriving packet. For a packet arriving at time n let $\tilde{X}_n$ be the number of packets it finds in the system and let J_n be the phase of the next packet arrival. Define $y_{i,j} = Pr\{\tilde{X}_n = i, J_n = j\}|_{n\to\infty}, \ \ i \geq 0, \ 1 \leq j \leq n_t$. Let $\mathbf{y}_i = [y_{i,1}, y_{i,2}, \cdots, y_{i,n_t}]$, then we have

$$\mathbf{y}_0 = \sigma[\boldsymbol{x}_0(\mathbf{t}\boldsymbol{\alpha}) + \boldsymbol{x}_1(\mathbf{t}\boldsymbol{\alpha})b], \tag{4.142}$$

$$\mathbf{y}_i = \sigma[\boldsymbol{x}_i(\mathbf{t}\boldsymbol{\alpha})\bar{b} + \boldsymbol{x}_{i+1}(\mathbf{t}\boldsymbol{\alpha})b], \ i \geq 1. \tag{4.143}$$

The paramter σ is a normalizing constant and is obtained as follows. Since we have $\sum_{i=0}^{\infty} \mathbf{y}_i \mathbf{1} = 1$, then we have

$$\sigma \sum_{i=0}^{\infty} \boldsymbol{x}_i (\mathbf{t}\boldsymbol{\alpha})\mathbf{1} = 1. \tag{4.144}$$

By the definition of $\boldsymbol{x}_i$ it is clear that $\mathbf{c} = \sum_{i=0}^{\infty} \boldsymbol{x}_i$ is a vector representing the phase of the PH renewal process describing the arrival process, i.e. it is equivalent to $\mathbf{c} = \mathbf{c}(T + \mathbf{t}\boldsymbol{\alpha})$ with $\mathbf{c}\mathbf{1} = 1$. Hence

$$\sigma\sum_{i=0}^{\infty}\boldsymbol{x}_i(\mathbf{t}\boldsymbol{\alpha})\mathbf{1} = \sigma\mathbf{c}(\mathbf{t}\boldsymbol{\alpha})\mathbf{1} = \sigma(\mathbf{ct}) = 1. \quad (4.145)$$

Hence we have

$$\sigma = \lambda^{-1}, \quad (4.146)$$

where λ is the mean arrival rate of the packets, with $\lambda^{-1} = \boldsymbol{\alpha}(I-T)^{-1}\mathbf{1}$ which was shown in Alfa [6] that $\mathbf{ct} = \lambda$.

Let the waiting time in the queue be W_q with $w_i^q = Pr\{W_q = i\}$, $i \geq 0$. Then we have

$$w_0^q = \mathbf{y}_0\mathbf{1}, \quad \text{and} \quad (4.147)$$

$$w_i^q = \sum_{v=1}^{i}\mathbf{y}_v\mathbf{1}\binom{i-1}{v-1}b^v(1-b)^{i-v}, \; i \geq 1. \quad (4.148)$$

4.9 PH/PH/1 Queues

Both the Geo/PH/1 and PH/Geo/1 queues are special cases of the PH/PH/1 queue. However, the PH/PH/1 is richer than either one of them in that it has more applications in many fields, especially telecommunications. It addition, it can be analyzed in several different ways depending on the structures of the underline PH distributions.

Consider a single server system with phase type arrivals characterized by $(\boldsymbol{\alpha}, T)$ of dimension n_t and service times of phase type distribution characterized by $(\boldsymbol{\beta}, S)$ of dimension n_s. The mean arrival rate $\lambda^{-1} = \alpha(I-T)^{-1}\boldsymbol{e}$ and the mean service rate $\mu^{-1} = \beta(I-S)^{-1}\boldsymbol{e}$. We study the DTMC $\{(X_n, J_n, K_n)\}, n \geq 0$, where at time n, X_n, J_n and K_n represent the number of items in the system, the phase of service for the item in service, and the phase of arrival, respectively. The state space can be arranged as $\{(0,k) \cup (i,j,k),\; i \geq 1, j = 1,2,\cdots,n_s; k = 1,2,\cdots,n_t\}$. The transition matrix for this DTMC is a QBD writen as

$$P = \begin{bmatrix} B & C & & & \\ E & A_1 & A_0 & & \\ & A_2 & A_1 & A_0 & \\ & & A_2 & A_1 & A_0 \\ & & & \ddots & \ddots & \ddots \end{bmatrix}, \quad (4.149)$$

where the matrices $B, C, E, A_v, v = 0,1,2$ have dimensions $n_t \times n_t$, $n_t \times n_s n_t$, $n_s n_t \times n_t$, $n_s n_t \times n_s n_t$, respectively. These matrices are given as

$$B = T,\; E = T \otimes \mathbf{s},\; C = (\mathbf{t}\boldsymbol{\alpha}) \otimes \boldsymbol{\beta}$$
$$A_0 = (\mathbf{t}\boldsymbol{\alpha}) \otimes S,\; A_1 = (\mathbf{t}\boldsymbol{\alpha}) \otimes (\mathbf{s}\boldsymbol{\beta}) + T \otimes S \text{ and } A_2 = T \otimes (\mathbf{s}\boldsymbol{\beta}).$$

As explained in Chapter 2, the operator $\otimes$ is the Kronecker product such that for two matrices U of dimension $n_a \times n_b$ and V of dimension $n_c \times n_d$ we obtain a matrix $W = U \otimes V$ of dimension $n_a n_c \times n_b n_d$ as

$$W = \begin{bmatrix} U_{1,1}V & U_{1,2}V & \cdots & U_{1,n_b}V \\ \vdots & \vdots & \vdots & \vdots \\ U_{n_a,1}V & U_{n,2}V & \cdots & U_{n_a,n_b}V \end{bmatrix}.$$

Observation: We notice that if we replace (α, T) with $(1, \bar{a})$ and (β, S) with $(1, \bar{b})$, we actually have the Markov chain of the Geo/Geo/1 system. It is easy to see that both the GI/Geo/1 and Geo/G/1 are also special cases of the PH/PH/1 if we let the general interarrival times for GI/Geo/1 and the general service times for the Geo/G/1 to have finite supports. In summary, the PH/PH/1 system is a generalization of several well-known single server queues, in discrete time. We will also show later that the GI/G/1 is also a special case of the PH/PH/1 system. So the PH/PH/1 system is a very important discrete-time single server queueing model.

We now want to solve for $\boldsymbol{x}$ where

$$\boldsymbol{x} = \boldsymbol{x}P, \ \ \boldsymbol{x}\mathbf{1} = 1,$$

and

$$\boldsymbol{x} = [\boldsymbol{x}_0, \ \boldsymbol{x}_1, \ \cdots, \ ], \ \boldsymbol{x}_i = [\boldsymbol{x}_{i,1}, \ \boldsymbol{x}_{i,2}, \ \cdots, \ \boldsymbol{x}_{i,n_s}], \ i \geq 1,$$

$$\boldsymbol{x}_{i,j} = [x_{i,j,1}, \ x_{i,j,2}, \ \cdots, \ x_{i,j,n_t}], \ \boldsymbol{x}_0 = [x_{0,1}, \ x_{0,2}, \ \cdots, \ x_{0,n_t}].$$

We know that there is an R matrix which is the minimal nonnegative solution to the matrix quadratic equation

$$R = A_0 + RA_1 + R^2 A_2.$$

After R is known then we have

$$\boldsymbol{x}_{i+1} = \boldsymbol{x}_i R, \ i \geq 1.$$

For most practical problems we need to compute R using one of the efficient techniques discussed in Chapter 2. However, we could try and exploit a structure of the matrices, if they have useful structures. The key ones that often arise in the case of the PH/PH/1 queue is that of some zero rows for A_0 or some zero columns for A_2. As we know, for any zero rows in A_0 the corresponding rows in R are zero. Similarly for any zero columns in A_2 the corresponding columns in G are zero.

The next step is to compute the boundary values $[\boldsymbol{x}_0, \ \boldsymbol{x}_1]$, which are based on

$$[\boldsymbol{x}_0, \ \boldsymbol{x}_1] = [\boldsymbol{x}_0, \ \boldsymbol{x}_1] \begin{bmatrix} B & C \\ E & A_1 + RA_2 \end{bmatrix}, \ [\boldsymbol{x}_0, \ \boldsymbol{x}_1]\mathbf{1} = 1.$$

This is then normalized as

$$\boldsymbol{x}_0\mathbf{1}+\boldsymbol{x}_1(I-R)^{-1}\mathbf{1}=1.$$

Several desired performance measures can then be computed accordingly.

4.9.1 Examples

4.9.1.1 A Numerical Example

Consider the case where

$$\boldsymbol{\alpha}=[1,\ 0,\ 0],\ T=\begin{bmatrix}.2 & .8 & 0\\ 0 & .3 & .7\\ 0 & 0 & 0.1\end{bmatrix},\ \boldsymbol{\beta}=[.25,\ \ .75],\ S=\begin{bmatrix}.8 & 0\\ 0 & .4\end{bmatrix}.$$

First we check the stability condition. We have

$$\lambda^{-1}=\boldsymbol{\alpha}(I-T)^{-1}\mathbf{1}=1/.2639,\ \mu^{-1}=\boldsymbol{\beta}(I-S)^{-1}\mathbf{1}=1/.4,$$

We obtain $\lambda/\mu=0.65975<1$. Hence the system is stable.

We calculate

$$A_0=(\mathbf{t}\boldsymbol{\alpha})\otimes S=\begin{bmatrix}0 & 0 & 0 & 0 & 0 & 0\\ 0 & 0 & 0 & 0 & 0 & 0\\ 0 & 0 & 0 & 0 & 0 & 0\\ 0 & 0 & 0 & 0 & 0 & 0\\ .72 & 0 & 0 & 0 & 0 & 0\\ 0 & .36 & 0 & 0 & 0 & 0\end{bmatrix}$$

$$A_1=T\otimes S+(\mathbf{t}\boldsymbol{\alpha})\otimes\mathbf{s}\boldsymbol{\beta}=\begin{bmatrix}.16 & 0 & .64 & 0 & 0 & 0\\ 0 & .08 & 0 & .32 & 0 & 0\\ 0 & 0 & .24 & 0 & .56 & 0\\ 0 & 0 & 0 & .12 & 0 & .28\\ .045 & .135 & 0 & 0 & .08 & 0\\ .135 & .405 & 0 & 0 & 0 & .04\end{bmatrix}$$

$$A_2=T\otimes(\mathbf{s}\boldsymbol{\beta})=\begin{bmatrix}.01 & .03 & .04 & .12 & 0 & 0\\ .03 & .09 & .12 & .36 & 0 & 0\\ 0 & 0 & .015 & .045 & .035 & .105\\ 0 & 0 & .045 & .135 & .105 & .315\\ 0 & 0 & 0 & 0 & .005 & .015\\ 0 & 0 & 0 & 0 & .015 & .045\end{bmatrix}.$$

For this example we have

$$R = \begin{bmatrix} 0 & 0 & 0 & 0 & 0 & 0 \\ 0 & 0 & 0 & 0 & 0 & 0 \\ 0 & 0 & 0 & 0 & 0 & 0 \\ 0 & 0 & 0 & 0 & 0 & 0 \\ .9416 & .2313 & .8829 & .3173 & .5874 & .2363 \\ .0103 & .4195 & .0141 & .1666 & .0105 & .0541 \end{bmatrix}.$$

If we write the matrix blocks A_v, $v = 0, 1, 2$ in smaller blocks such that $A^v_{i,j}$, $i, j = 1, 2, 3$ is the $(i, j)^{th}$ block of the matrix A_v we find that

$$A_0 = \begin{bmatrix} 0 & 0 & 0 \\ 0 & 0 & 0 \\ A^0_{3,1} & 0 & 0 \end{bmatrix}, A_1 = \begin{bmatrix} A^1_{1,1} & A^1_{1,2} & 0 \\ 0 & A^1_{2,2} & A^1_{2,3} \\ A^1_{3,1} & 0 & A^1_{3,3} \end{bmatrix}, A_2 = \begin{bmatrix} A^2_{1,1} & A^2_{1,2} & 0 \\ 0 & A^2_{2,2} & A^2_{2,3} \\ 0 & 0 & A^2_{3,3} \end{bmatrix}.$$

With the structure A_0 we know that our R matrix is of the form $R = \begin{bmatrix} 0 & 0 & 0 \\ 0 & 0 & 0 \\ R_{3,1} & R_{3,2} & R_{3,3} \end{bmatrix}$. This is one example where applying the linear approach for computing R may be beneficial because we only need to compute the three blocks of R, and this could reduce computational efforts compared to the quadratic algorithms in some instances. In this particular case we have

$$R_{3,1} = A^0_{3,1} + R_{3,1}A^1_{1,1} + R_{3,3}A^1_{3,1} + R_{3,3}R_{3,1}A^2_{1,1}, \tag{4.150}$$

$$R_{3,2} = R_{3,1}A^1_{1,2} + R_{3,2}A^1_{2,2} + R_{3,3}R_{3,1}A^2_{1,2} + R_{3,3}R_{3,2}A^2_{2,2}, \tag{4.151}$$

$$R_{3,3} = R_{3,2}A^1_{2,3} + R_{3,3}A^1_{3,3} + R_{3,3}R_{3,2}A^2_{2,3} + R_{3,3}R_{3,3}A^2_{3,3}. \tag{4.152}$$

Writing these equations simply as $R_{3,i} = f_i(R, A_0, A_1, A_2), i = 1, 2, 3$, we can now write an iterative process given as

$$R_{3,i}(k+1) = f_i(R(k), A_0, A_1, A_2). \tag{4.153}$$

What we now do is set $R_{3,i}(0) := 0$ and then apply an iterative process until $|R_{3,i}(k+1) - R_{3,i}(k)|_{u,v} < \varepsilon$, $(u, v) = 1, 2, 3$, $\forall i$, where ε is a very small number, usually 10^{-12} is acceptable.

Sometimes we find that by re-arranging the state space we can find useful structures.

4.9.1.2 Another Example

Consider a system with Negative binomial arrival process which can be represented by a phase type distribution (α, T) with

$$\boldsymbol{\alpha} = [1,\ 0,\ 0],\ \ T = \begin{bmatrix} 1-a & a & 0 \\ 0 & 1-a & a \\ 0 & 0 & 1-a \end{bmatrix},$$

and Negative binomial service process which is represented by $(\boldsymbol{\beta}, S)$ with

$$\boldsymbol{\beta} = [1,\ 0],\ \ S = \begin{bmatrix} 1-b & b \\ 0 & 1-b \end{bmatrix}.$$

Here we can see that there are different advantages with arranging our DTMC as i) $\{(X_n, J_n, K_n)\}, n \geq 0$, or as ii) $\{(X_n, K_n, J_n)\}, n \geq 0$. In the first arrangement we have

$$A_0 = S \otimes (\mathbf{t}\boldsymbol{\alpha}),\ \ \text{and}\ \ A_2 = (\mathbf{s}\boldsymbol{\beta}) \otimes T.$$

With this arrangement, our A_2 has only one non-zero column and one zero-column. In fact, if our Negative binomial for service was of higher dimension $n_s > 2$ we would have $n_s - 1$ zero columns. In this case we are better to study the G matrix because this matrix has equivalent zero columns. We can then obtain matrix R from matrix G by using the formula of Equation (2.100).

If however, we use the second arrangement then we have

$$A_0 = (\mathbf{t}\boldsymbol{\alpha}) \otimes S,\ \ \text{and}\ \ A_2 = T \otimes (\mathbf{s}\boldsymbol{\beta}).$$

With this arrangement, our A_0 has two zero rows. Again had our Negative binomial for arrival process been of order $n_t > 3$ we would have $n_t - 1$ zero columns. This would result in corresponding $n_t - 1$ zero rows of the matrix R. So here we are better to study the matrix R directly.

The standard methods presented are now used to obtain the performance measures of this system after matrix R is obtained.

In the end we may not have zero rows for R or zero columns for G, but the matrices A_v, $v = 0, 1, 2$ may be very sparse, thereby making it easier to work in smaller blocks of matrices. This sometimes assists in reducing computational efforts required for computing R and G.

The mean waiting time in the system μ_W is given by Little's Law as

$$\mu_L = \lambda \mu_W. \tag{4.154}$$

4.9.2 Waiting Time Distirbution

The waiting time distribution can be studied in at least three different ways. All the three methods will be presented here. It goes without saying that any of the methods may also be used for all the single node queues that have been presented so far since they are all special cases of this PH/PH/1 system.

In order to obtain the waiting time distribution for a packet, irrespective of the method used, we first have to determine the state of the system as seen by an arriving customer. Let us define $\mathbf{y}$ as the steady state vector describing the state of the system as seen by an arriving customer, with $\mathbf{y} = [\mathbf{y}_0, \mathbf{y}_1, \ldots]$. Then we have

$$\mathbf{y}_0 = \lambda^{-1}[x_0(\mathbf{t}\boldsymbol{\alpha}) + x_1(\mathbf{t}\boldsymbol{\alpha} \otimes \mathbf{s})], \tag{4.155}$$

$$\mathbf{y}_i = \lambda^{-1}[x_i(\mathbf{t}\boldsymbol{\alpha}) \otimes S) + x_{i+1}(\mathbf{t}\boldsymbol{\alpha} \otimes \mathbf{s}\boldsymbol{\beta})],\ i \geq 1. \tag{4.156}$$

By letting $F = \lambda^{-1}[(\mathbf{t}\boldsymbol{\alpha}) \otimes S + R((\mathbf{t}\boldsymbol{\alpha}) \otimes (\mathbf{s}\boldsymbol{\beta}))]$, we can write

$$\mathbf{y}_i = \boldsymbol{x}_1 R^{i-1} F,\ i \geq 1.$$

Let W^q be the waiting time in the queue, with $w_j^q = Pr\{W^q = j\}$ and $\tilde{W}_j^q = Pr\{W^q \geq j\}$.

Method One:

Further using B_r^k as defined in Section 4.7.1
then we have

$$w_0^q = \mathbf{y}_0 \mathbf{1} \tag{4.157}$$

$$w_i^q = \sum_{k=1}^{i} y_k (\mathbf{1} \otimes I) B_i^k \mathbf{1},\ i \geq 1. \tag{4.158}$$

Method Two:

Another approach that can be used for studying the waiting time is as follows. An arriving packet will wait for at least k units of time if at its arrival it finds $i \geq 1$ packets ahead of it in the system and the server takes no less than $k \geq 1$ units of time to serve these i packets, i.e. no more than $i-1$ service completions in the interval $(0, k-1)$. Let $U(i,n)$ be the probability that the number of packets served in the interval $(0,n)$ is i. Then we have

$$\tilde{W}_k^q = \sum_{i=1}^{\infty} \mathbf{y}_i (\mathbf{1}_{n_t} \otimes I_{n_s}) [\sum_{j=0}^{i-1} U(j, k-1)] \mathbf{1}$$

$$= \sum_{i=1}^{\infty} \boldsymbol{x}_1 R^{i-1} F (\mathbf{1}_{n_t} \otimes I_{n_s}) [\sum_{j=0}^{i-1} U(j, k-1)] \mathbf{1},$$

which results in

$$\tilde{W}_k^q = \boldsymbol{x}_1 (I - R)^{i-1} [\sum_{j=0}^{k-1} R^j F (\mathbf{1}_{n_t} \otimes I_{n_s}) U(j, k-1)] \mathbf{1}. \tag{4.159}$$

Method Three:

A third approach uses the absorbing Markov chain idea. First we let H be the phase of arrival after the target packet has arrived, and define $W_{i,j}^q = Pr\{W^q \leq i, H = j\}$, $i \geq 0$, $1 \leq j \leq n_t$. Define a transition matrix $\hat{P}$ of an absorbing Markov chain as

$$\hat{P} = \begin{bmatrix} I & & & & \\ I \otimes \mathbf{s} & I \otimes S & & & \\ & I \otimes (\mathbf{s}\boldsymbol{\beta}) & I \otimes S & & \\ & & I \otimes (\mathbf{s}\boldsymbol{\beta}) & I \otimes S & \\ & & & \ddots & \ddots \end{bmatrix}. \tag{4.160}$$

Define $\mathbf{z}^{(0)} = \mathbf{y}$, with $\mathbf{z}_i^{(0)} = \mathbf{y}_i$. Further write $\mathbf{W}^q = [W_1^q, W_2^q, \cdots]$ and $W_i^q = [W_{i,1}^q, W_{i,2}^q, \cdots, W_{i,n_t}^q]$, then we have

$$\mathbf{z}^{(i)} = \mathbf{z}^{(i-1)}\hat{P} = \mathbf{z}^{(0)}\hat{P}^i, \; i \geq 1, \tag{4.161}$$

with

$$W_i^q = \mathbf{z}_0^{(i)}, \tag{4.162}$$

which is the waiting time vector that also considers the phase of arrival.

The busy period of the PH/PH/1 queue can be obtained as a special case of the busy period of the MAP/PH/1 queue as shown in Frigui and Alfa [41].

4.9.2.1 Workload

Consider the DTMC $(V_n, K_n), n \geq 0; V_n \geq 0, K_n = 1, 2, \cdots, n_t$, where at time n, V_n is the workload in the system and K_n is the phase of the interarrival time. Let the probability transition matrix of this DTMC be P. Then P is given as

$$P = \begin{bmatrix} C_0 & C_1 & C_2 & C_3 & \cdots \\ C_0 & C_1 & C_2 & C_3 & \cdots \\ & C_0 & C_1 & C_2 & \cdots \\ & & C_0 & C_1 & \cdots \\ & & & \ddots & \ddots \end{bmatrix},$$

where

$$C_0 = T \otimes I, C_k = (\mathbf{t}\boldsymbol{\beta}) \otimes S^{k-1}\mathbf{s}, \; k \geq 1.$$

This workload model has the same structure as the M/G/1 type markov chains discussed in Chapter 2. In addition the boundary has a very special structure similar to that of the workload for the Geo/G/1 system as expected. So we can go ahead and apply the M/G/1 type results here and also try and capitalize on the associated special structure at the boundary.

4.9.2.2 Age Process

Here we present the age proces model which studies the age of the leading packet in a FCFS system. Consider the DTMC represented by $(Y_n, J_n, K_n), n \geq 0; Y_n \geq 0, K_n =$

$1,2,\cdots,n_t;J_n=1,2,\cdots,n_s$. Here Y_n is the age at time n, and K_n and J_n are the phases of arrivals and service, respectively. The associated probability transition matrix P has the following form

$$P=\begin{bmatrix} C_0 & C_1 & & & & \\ B_1 & A_1 & A_0 & & & \\ B_2 & A_2 & A_1 & A_0 & & \\ B_3 & A_3 & A_2 & A_1 & A_0 & \\ \vdots & \vdots & \vdots & \vdots & \vdots & \ddots \end{bmatrix},$$

where

$$C_0=T,\ C_1=(\mathbf{t}\boldsymbol{\alpha})\otimes\boldsymbol{\beta},\ B_k=\mathbf{s}T^k, k\geq 1,$$

$$A_0=S\otimes I,\ A_j=(\mathbf{s}\boldsymbol{\beta})\otimes T^{j-1}(\mathbf{t}\boldsymbol{\alpha}),\ j\geq 1.$$

It is immediately clear that this has the GI/M/1 structure and the results from that class of Markov chains can be used to analyze this system. If the system is stable then from the steady state results of the DTMC one can obtain several performance measures such as the waiting time distribution and the distribution of the number in the system. For a detailed treatment of such systems see [100].

4.9.3 PH/PH/1 Queues at points of events

We study this system at epochs of events. By that we mean at epochs of an arrival, a departure, or of joint arrival and departure. The idea was first studied by Latouche and Ramaswami [66] for the continuous case of the PH/PH/1. By studying it at points of events we end up cutting down on the state space requirements, and hence the computational effort required at the iteration stage, even though the front-end work does increase substantially.

Let $\tau_j,\ j\geq 0$ be a collection of the epochs of events. Consider the process $\{(N_j,I_j,J_j),j\geq 0\}$, where $N_j=N(\tau_j+)$ is the number in the system just after the jth epoch, I_j the indicator equal to $+$ if τ_j is an arrival, $-$ if τ_j is a departure, and $*$ if τ_j is a joint arrival and departure. It is immediately clear that if τ_j is an arrival then J_j is the phase of service in progress at time τ_j and similarly if τ_j is a departure epoch then J_j is the phase of the interarrival time. Of course, when τ_j is an epoch of joint arrival and departure then both service (in case of a customer in the system) and arrival processes are both initialized. The process $\{(N_j,I_j,J_j),j\geq 0\}$ is an embedded Markov chain with its transition matrix P given as

$$P=\begin{bmatrix} \mathbf{0} & C & & & \\ E & A_1 & A_0 & & \\ & A_2 & A_1 & A_0 & \\ & & A_2 & A_1 & A_0 \\ & & & \ddots & \ddots & \ddots \end{bmatrix}, \quad (4.163)$$

where

$$A_0 = \begin{bmatrix} A_0^{++} & \mathbf{0} & \mathbf{0} \\ A_0^{*+} & \mathbf{0} & \mathbf{0} \\ A_0^{-+} & \mathbf{0} & \mathbf{0} \end{bmatrix}, \quad A_2 = \begin{bmatrix} \mathbf{0} & \mathbf{0} & A_2^{+-} \\ \mathbf{0} & \mathbf{0} & A_2^{*-} \\ \mathbf{0} & \mathbf{0} & A_2^{--} \end{bmatrix},$$

$$A_1 = \begin{bmatrix} \mathbf{0} & A_1^{+*} & \mathbf{0} \\ \mathbf{0} & A_1^{**} & \mathbf{0} \\ \mathbf{0} & A_1^{-*} & \mathbf{0} \end{bmatrix}, \quad E = \begin{bmatrix} A_2^{+-} \\ A_2^{*-} \\ A_2^{--} \end{bmatrix}, \quad C = \begin{bmatrix} \mathbf{1} \boldsymbol{\beta} \, \mathbf{0} \, \mathbf{0} \end{bmatrix}.$$

All the matrices A_i, $i = 0,1,2$ are of order $(n_s + n_t + 1) \times (n_s + n_t + 1)$, matrix E is of order $(n_s + n_t + 1) \times n_t$, and matrix C is of order $n_t \times (n_s + n_t + 1)$. The block elements are given as follows:

- $A_0^{++} = \sum_{k=0}^{min(n_s,n_t)} (\boldsymbol{\alpha} T^k \mathbf{t}) S^{k+1} = \sum_{k=0}^{min(n_s,n_t)} a_{k+1} S^{k+1}$, of order $n_s \times n_s$
- $A_0^{*+} = \sum_{k=0}^{min(n_s,n_t)} (\boldsymbol{\alpha} T^k \mathbf{t}) \boldsymbol{\beta} S^{k+1} = \sum_{k=0}^{min(n_s,n_t)} a_{k+1} \boldsymbol{\beta} S^{k+1}$, of order $1 \times n_s$
- $A_0^{-+} = \sum_{k=0}^{min(n_s,n_t)} (T^k \mathbf{t})(\boldsymbol{\beta} S^{k+1})$, of order $n_t \times n_s$
- $A_2^{+-} = \sum_{k=0}^{min(n_s,n_t)} (S^k \mathbf{s})(\boldsymbol{\alpha} T^{k+1})$, of order $n_s \times n_t$
- $A_2^{*-} = \sum_{k=0}^{min(n_s,n_t)} (\boldsymbol{\beta} S^k \mathbf{s}) \boldsymbol{\alpha} T^{k+1} = \sum_{k=0}^{min(n_s,n_t)} s_{k+1} \boldsymbol{\alpha} T^{k+1}$, of order $1 \times n_t$
- $A_2^{--} = \sum_{k=0}^{min(n_s,n_t)} (\boldsymbol{\beta} S^k \mathbf{s}) T^{k+1} = \sum_{k=0}^{min(m,n)} s_{k+1} T^{k+1}$, of order $n_t \times n_t$
- $A_1^{+*} = \sum_{k=0}^{min(n_s,n_t)} (\boldsymbol{\alpha} T^k \mathbf{t}) S^k \mathbf{s} = \sum_{k=0}^{min(n_s,n_t)} a_{k+1} S^k \mathbf{s}$, of order $n_s \times 1$
- $A_1^{**} = \sum_{k=0}^{min(n_s,n_t)} (\boldsymbol{\alpha} T^k \mathbf{t})(\boldsymbol{\beta} S^k \mathbf{s}) = \sum_{k=0}^{min(n_s,n_t)} a_{k+1} s_{k+1}$, of order 1×1
- $A_1^{-*} = \sum_{k=0}^{min(n_s,n_t)} (\boldsymbol{\beta} S^k \mathbf{s}) T^k \mathbf{t} = \sum_{k=0}^{min(n_s,n_t)} s_{k+1} (T^k \mathbf{t})$, of order $n_t \times 1$

We later show how to simplify these block matrices for computational purposes. The type of simplification approach adopted depends on whether one uses the remaining time or elapsed time approach of representation for the interarrival and service processes.

The stationary vector $\boldsymbol{x}$ is given as

$$\boldsymbol{x} = \boldsymbol{x} P, \quad \boldsymbol{x} \mathbf{1} = 1,$$

with

$$\boldsymbol{x} = [\boldsymbol{x}_0, \, \boldsymbol{x}_1, \, \boldsymbol{x}_2, \, \cdots], \quad \boldsymbol{x}_i = [\boldsymbol{x}_i^+, \, x_i^*, \, \boldsymbol{x}_i^-], \; i \geq 1, \text{ and } \boldsymbol{x}_0 = \boldsymbol{x}_0^-.$$

The matrix R and the stationary distributions are given as

$$R = A_0 + RA_1 + R^2 A_2, \quad \text{and} \quad \boldsymbol{x}_{i+1} = \boldsymbol{x}_i R, \; i \geq 1.$$

We also know that there is a matrix G which is the minimal non-negative solution to the matrix quadratic equation

$$G = A_2 + A_1 G + A_0 G^2, \tag{4.164}$$

and the relationship between R and G is such that

$$R = A_0 (I - A_1 - A_0 G)^{-1}. \tag{4.165}$$

Based on the structure of A_2 we know that the matrix G is of the form

$$G = \begin{bmatrix} \mathbf{0} & \mathbf{0} & G^{+-} \\ \mathbf{0} & 0 & G^{*-} \\ \mathbf{0} & \mathbf{0} & G^{--} \end{bmatrix},$$

where

$$G^{+-} = A_2^{+-} + A_1^{+*}G^{*-} + A_0^{++}G^{+-}G^{--}, \tag{4.166}$$

$$G^{*-} = A_2^{*-} + A_1^{**}G^{*-} + A_0^{*+}G^{+-}G^{--}, \tag{4.167}$$

and

$$G^{--} = A_2^{--} + A_1^{-*}G^{*-} + A_0^{-+}G^{+-}G^{--}. \tag{4.168}$$

We write $\hat{G} = G^{+-}G^{--}$, and $H = (1 - A_1^{**})^{-1}(A_2^{*-} + A_0^{*+}\hat{G})$. By manipulating Equations (4.166) to (4.168) we have

$$\hat{G} = A_2^{+-}(A_2^{--} + A_1^{-*}H + A_1^{-+}\hat{G}) + (A_1^{+*}H + A_0^{++}\hat{G})(A_2^{--} + A_1^{-*}H + A_0^{-+}\hat{G}). \tag{4.169}$$

This is a matrix equation which can be applied iteratively to compute $\hat{G}$ which is of order $m \times n$. After computing $\hat{G}$ we can then obtain the block matrices of G as follows:

$$G^{*-} = H. \tag{4.170}$$

$$G^{--} = A_2^{--} + A_1^{-*}H + A_0^{-+}\hat{G}. \tag{4.171}$$

$$G^{+-} = A_2^{+-} + A_1^{+*}H + A_0^{++}\hat{G}. \tag{4.172}$$

So it might be easier to compute the matrix G first and then compute R from it because computing R from Equation (4.165) only requires us to know the first block of row of the inverse matrix as a result of the structure of matrix A_0 which has only the block column that is non-zero.

Let us denote $D = (I - A_1 - A_0G)$. It is clear that the structure of D is of the form

$$D = \begin{bmatrix} D_{11} & D_{12} & D_{13} \\ \mathbf{0} & D_{22} & D_{23} \\ \mathbf{0} & D_{32} & D_{33} \end{bmatrix}. \tag{4.173}$$

It's inverse F such that $FD = I$ can be written in block form as

$$F = \begin{bmatrix} F_{11} & F_{12} & F_{13} \\ F_{21} & F_{22} & F_{23} \\ F_{31} & F_{32} & F_{33} \end{bmatrix}. \tag{4.174}$$

After simple algebraic calculations we have

$$F_{11} = D_{11}^{-1}, \tag{4.175}$$

$$F_{12} = [D_{11}^{-1}D_{13}D_{33}^{-1}D_{32} - D_{11}^{-1}D_{12}][D_{22} - D_{23}D_{33}^{-1}D_{32}]^{-1}, \tag{4.176}$$

and

$$F_{13} = -(F_{12}D_{23} + D_{11}^{-1}D_{13})D_{33}^{-1}. \tag{4.177}$$

From here we know that

$$R = \begin{bmatrix} A_0^{++}F_{11} & A_0^{++}F_{12} & A_0^{++}F_{13} \\ A_0^{*+}F_{11} & A_0^{*+}F_{12} & A_0^{*+}F_{13} \\ A_0^{-+}F_{11} & A_0^{-+}F_{12} & A_0^{-+}F_{13} \end{bmatrix}. \tag{4.178}$$

So the computation of the matrix R involves iterative computations of $n_s \times n_t$ matrix $\hat{G}$ from which the three block matrices G^{+-}, G^{*-} and G^{--} are directly computed, and of partial computation of the inverse matrix F, i.e. we only compute the first block row of this matrix. Computing the R matrix is thus reasonably efficient.

Consider the U matrix associated with this Markov chain, with $U = A_1 + A_0 G = A_1 + RA_2$, we have U given as

$$U = \begin{bmatrix} \mathbf{0} & U^{+*} & U^{+-} \\ \mathbf{0} & U^{**} & U^{*-} \\ \mathbf{0} & U^{-*} & U^{--} \end{bmatrix}. \tag{4.179}$$

Then we have

$$U^{+*} = A^{+*},\; U^{**} = A^{**},\; U^{-*} = A^{-*} \tag{4.180}$$

and

$$U^{+-} = A^{++}G^{+-},\; U^{*+} = A^{*+}G^{*-},\; U^{--} = A^{-+}G^{--}. \tag{4.181}$$

It is clear that because of the structures of T and S we know that $T^j = 0,\ \forall j \geq n$ and $S^i = 0,\ \forall i \geq m$. Let us assume that for both the interarrival and service times we adopt the remaining time approach for representation. Let us define the following:

- $I(k,j)$ as a $k \times k$ matrix with zero elements and 1 in locations $(j,1)$, $(j+1,2),\cdots,(k,j-k+1)$. It is immediately clear that the identity matrix I_k of order k can be written as $I_k = I(k,0)$. The expression $S^k = I(n_s,k+1)$ and $T^k = I(n_t,k+1)$. Also $I(k,j)I(k,1) = I(k,j+1)$.
- $b_i = \sum_{j=1}^{min(n_t,n_s-i)} a_j s_{i+j}$.
- $d_i = \sum_{j=1}^{min(n_s,n_t-i)} s_j a_{i+j}$.

Then we have

1. $A_0^{++} = \sum_{k=0}^{min(n_s,n_t)} a_{k+1} I(n_s,k+1) = \begin{bmatrix} 0 & 0 & 0 & \cdots & 0 \\ a_1 & 0 & 0 & \cdots & 0 \\ a_2 & a_1 & 0 & \cdots & 0 \\ a_3 & a_2 & a_1 & \cdots & 0 \\ \vdots & \ddots & \ddots & \ddots & \vdots \end{bmatrix}_{n_s \times n_s}$
2. $A_0^{*+} = \sum_{k=0}^{min(n_s,n_t)} a_k \beta S^{k+1} = [b_1,\ b_2,\ \cdots,\ b_{n_s}]$

3. $A_0^{-+} = \sum_{k=0}^{min(n_s,n_t)} (T^k \mathbf{t})(\beta S^{k+1}) = \begin{bmatrix} s_2 & s_3 & s_4 & \cdots & s_{n_s-2} & s_{n_s-1} & s_{n_s} & 0 \\ s_3 & s_4 & s_5 & \cdots & s_{n_s-1} & s_{n_s} & 0 & 0 \\ s_4 & s_5 & s_6 & \cdots & s_{n_s} & 0 & 0 \\ \vdots & \vdots & \vdots & \vdots & \vdots & \vdots & \vdots & \vdots \end{bmatrix}_{n_t \times n_s}$

4. $A_2^{+-} = \sum_{k=0}^{min(n_s,n_t)} (S^k \mathbf{s})(\alpha T^{k+1}) = \begin{bmatrix} a_2 & a_3 & a_4 & \cdots & a_{n-2} & a_{n-1} & a_n & 0 \\ a_3 & a_4 & a_5 & \cdots & a_{n-1} & a_n & 0 & 0 \\ a_4 & a_5 & a_6 & \cdots & a_n & 0 & 0 \\ \vdots & \vdots & \vdots & \vdots & \vdots & \vdots & \vdots & \vdots \end{bmatrix}_{n_s \times n_t}$

5. $A_2^{*-} = \sum_{k=0}^{min(n_s,n_t)} s_{k+1}(\alpha T^{k+1}) = [d_1,\ d_2,\ \cdots,\ d_{n_t}]$

6. $A_2^{--} = \sum_{k=0}^{min(n_s,n_t)} s_{k+1} I(n_t, k+1) = \begin{bmatrix} 0 & 0 & 0 & \cdots & 0 \\ s_1 & 0 & 0 & \cdots & 0 \\ s_2 & s_1 & 0 & \cdots & 0 \\ s_3 & s_2 & s_1 & \cdots & 0 \\ \vdots & \ddots & \ddots & \ddots & \vdots \end{bmatrix}_{n_t \times n_t}$

7. $A_1^{+*} = \sum_{k=0}^{min(n_s,n_t)} t_{k+1} S^k \mathbf{s} = \begin{bmatrix} a_1 \\ a_2 \\ \vdots \\ a_{min(n_s,n_t)} \\ 0 \\ \vdots \\ 0 \end{bmatrix}_{n_s \times 1}$

8. $A_1^{**} = \sum_{k=0}^{min(n_s,n_t)} (\alpha T^k \mathbf{t})(\beta S^k \mathbf{s}) = \sum_{k=0}^{min(n_s,n_t)} a_k s_k$

9. $A_1^{-*} = \sum_{k=0}^{min(n_s,n_t)} s_{k+1}(T^k \mathbf{t}) = \begin{bmatrix} s_1 \\ s_2 \\ \vdots \\ s_{min(n_s,n_t)} \\ 0 \\ \vdots \\ 0 \end{bmatrix}_{n_t \times 1}$

It is clear from Latouche and Ramaswami [66] that it is straightforward to obtain the stationary distributions at arrivals and also at departures from the information about the system at epoch of events. They also show how to obtain the information at arbitrary times. By applying similar arguments used by Latouche and Ramaswami [66] we can also derive the stationary distributions at arbitrary times from the results of the case of time of events. It is however not necessary to present those results here since the techniques used will practically be the same. But we point out that if the interest is to obtain the system performance at arbitrary times, then it is more efficient to go ahead and apply the results of Alfa and Xue [3] directly, rather than try to study the system at epochs of events and then recover the arbitrary times results from it.

4.10 GI/G/1 Queues

The GI/G/1 is the most basic and the most general of single server queues. However, obtaining an exact solution for its stationary distribution of number in the system has been elusive. We have to turn to numerical approaches. Here we present a very efficient algorithmic approach for studying this system of queues.

We let the arrival process be described as follows. Let the interarrival times between two consecutive packets be $\mathscr{A} > 0$, and define $a_k = Pr\{\mathscr{A} = k\}$. Similarly we let the service time of each packet be $\mathscr{S} > 0$ and define $b_k = Pr\{\mathscr{S} = k\}$. We assume the interarrival and service times are independent of each other. It was pointed out in Alfa [6] and earlier in Neuts [84] that any discrete probability distribution with finite support can be represented as a PH distribution. Alfa [6] also pointed out that even if the support is infinite we can still represent it by the IPH distribution, which is just a PH distribution with infinite states. Hence GI/G/1 system with finite support for interarrival and service times can be modelled as a PH/PH/1 system. It was shown in Section 3.3.4 how to represent the general distribution as a PH distribution. In this section we focus mainly on the case where both the interarrival and service times have finite supports.

Consider the case where $a_k = 0, \forall k > K_a = n_t < \infty$, and $s_j = 0, \forall j > K_s = n_s < \infty$. We further write

$$\mathbf{a} = [a_1,\ a_2,\ \cdots,\ a_{n_t}],\ \ \mathbf{b} = [b_1,\ b_2,\ \cdots,\ b_{n_s}].$$

There are two ways of representing each of these two distributions as PH distributions, one is using the elapsed time (ET) approach and the other is using the remaining time (RT) approach. This implies that we can actually model the GI/G/1 system in four different ways as PH/PH/1 system as follows. We write (X,Y) and let X represent interarrival times type and Y service times type, then we have four possible selections from: (ET,ET), (RT,RT), (ET,RT) and (RT,ET). We will present only two of these and let the reader develop the other two. We discuss (RT,RT) and (ET,ET).

4.10.1 The (RT,RT) representation for the GI/G/1 queue

Using the results of Section 3.3.4 we can write the interarrival times and service times as RT PH distributions as follows. For the interarrival time with PH written as (α, T) we have

$$\alpha = [a_1,\ a_2,\ a_3,\ \cdots,\ a_{n_t}],\ T = \begin{bmatrix} \mathbf{0}_{n_t-1} & 0 \\ I_{n_t-1} & \mathbf{0}^T_{n_t-1} \end{bmatrix},$$

where $\mathbf{0}(j)$ is a j row of zeros and $I(k)$ is an identity matrix of order k. Similarly we can write the PH distribution for the service times as (β, S) with

$$\alpha = [b_1,\ b_2,\ b_3,\ \cdots,\ b_{n_s}],\ S = \begin{bmatrix} \mathbf{0}_{n_s-1} & 0 \\ I_{n_s-1} & \mathbf{0}^T_{n_s-1} \end{bmatrix}.$$

We consider a trivariate Markov chain $\{X_n, K_n, J_n\}$, where at time n, X_n is the number of packets in the system, K_n the remaining time to next arrival and J_n the remaining service time of the packet in service. The DTMC representing the number in the system for this GI/G/1 queue is $\Delta = \{(0,k) \cup (i,k,j)\}; k = 1,2,\cdots,n_t; j = 1,2,\cdots,n_s; i \geq 1$, i.e. we have number in the system, phase of arrival and phase of service – in that order. The transition matrix for this system can be written as QBD as

$$P = \begin{bmatrix} B & C & & & \\ E & A_1 & A_0 & & \\ & A_2 & A_1 & A_0 & \\ & & \ddots & \ddots & \ddots \end{bmatrix},$$

with

$$B = T,\ C = (\mathbf{t}\alpha) \otimes \beta,\ E = T \otimes \mathbf{s},\ A_0 = (\mathbf{t}\alpha) \otimes S,$$
$$A_1 = (\mathbf{t}\alpha) \otimes (\mathbf{s}\beta) + T \otimes S,\ A_2 = T \otimes (\mathbf{s} \otimes \beta).$$

It is immediately clear that we can apply the matrix-geometric results to this system, in fact we can use the results from the PH/PH/1 case here directly. The special feature of this model is that we have additional structures which we can exploit regarding the block matrices. The first one is that because of our arrangement of the state space we find that our block matrix A_0 has only one block of rows that is non-zero, as such the structure of our R matrix is

$$R = \begin{bmatrix} R_1 & R_2 & R_3 & \cdots & R_{n_t} \\ \mathbf{0} & \mathbf{0} & \mathbf{0} & \cdots & \mathbf{0} \\ \mathbf{0} & \mathbf{0} & \mathbf{0} & \cdots & \mathbf{0} \\ \vdots & \vdots & \vdots & \cdots & \mathbf{0} \\ \mathbf{0} & \mathbf{0} & \mathbf{0} & \cdots & \mathbf{0} \end{bmatrix}. \tag{4.182}$$

Hence we can apply the standard linear algorithm for computing matrix R so as to work in smaller blocks. It is probably easier to work with than any of the quadratically convergent algorithms in this special case. Based on this we have the following equations

$$R_k = a_k S + R_1 a_k(\mathbf{s}\beta) + R_{k+1} S + R_1 R_{k+1}(\mathbf{s}\beta),\ \ 1 \leq k \leq n_t - 1, \tag{4.183}$$

$$R_{n_t} = a_{n_t} S + R_1 a_{n_t} (\mathbf{s}\beta). \tag{4.184}$$

Let $C_0 = S$ and $C_1 = (\mathbf{s}\beta)$, and let us write

$$\mathscr{J} = C_0 \otimes I + C_1 \otimes R_1, \;\; \mathscr{H} = vecC_0 + (C_1^T \otimes I)vecR_1, \tag{4.185}$$

then we have

$$vecR_{n_t} = a_{n_t} \mathscr{H}, \tag{4.186}$$

and by back substitution we have

$$vecR_{n_t - k} = \sum_{v=0}^{k} a_{n_t - v} \mathscr{J}^{k-v} \mathscr{H}, \; k = 0,1,2,\cdots,n_t - 1. \tag{4.187}$$

All we really need to do is compute R_1 iteratively from

$$vecR_1 = \sum_{v=0}^{k} a_{n_t - 1} \mathscr{J}^{n_t - v - 1} \mathscr{H}, \tag{4.188}$$

starting with $R_1 := 0$ and the remaining $R_k,\; k = 2,3,\cdots,n_t$ are computed directly.

After the matrix R has been computed we can then obtain the boundary equations and all the information about the queue.

As a reminder from Section 2.5.2, for a matrix $A = [A_1,\, A_2,\, \cdots,\, A_N]$, where A_v is its vth column, we define $vecA = [A_1^T,\, A_2^T,\, \cdots,\, A_v,\, \cdots,\, A_N^T]^T$.

4.10.2 New algorithm for the GI/G/1 system

Alternatively we may arrange the state space as $\Delta = \{(0,k) \cup (i,j,k)\}; k = 1,2,\cdots,n_t; j = 1,2,\cdots,n_s; i \geq 1$, in which case our block matrices will be of the form

$$B = T,\; C = \beta \otimes (\mathbf{t}\alpha),\; E = \mathbf{s} \otimes T,\; A_0 = S \otimes (\mathbf{t}\alpha),$$

$$A_1 = (\mathbf{s}\beta) \otimes (\mathbf{t}\alpha) + S \otimes T,\; A_2 = (\mathbf{s} \otimes \beta) \otimes T.$$

With this arrangement we can exploit a different aspect of the matrices as follows. An efficient algorithm was developed by Alfa and Xue [3] for analysing this type of system. As usual we still need to compute the R matrix given as

$$R = A_0 + RA_1 + R^2 A_2.$$

It is clear that we can write

$$\boldsymbol{x}_0 (B + VE) = \boldsymbol{x}_0, \tag{4.189}$$

$$\boldsymbol{x}_0 \mathbf{1} + \boldsymbol{x}_0 V (I - R)^{-1} \mathbf{1} = 1, \tag{4.190}$$

where

$$V = B(I - A_1 - RA_2)^{-1} \text{ and} \tag{4.191}$$

$$\boldsymbol{x}_1 = \boldsymbol{x}_0 V. \tag{4.192}$$

A simple iteration method for computing R is

$$R(k+1) := A_0(I - A_1 - R(k)A_2)^{-1}, \tag{4.193}$$

with

$$R(0) := 0. \tag{4.194}$$

Even though this iteration method may not be efficient under normal circumstances, however for this particular problem it is very efficient once we capitalize on the structure of the problem. The Logarithmic Reduction method by Latouche and Ramaswami [65] is very efficient but only more efficient than this new linear algorithm at very high traffic cases, otherwise this algorithm is more efficient for medium to low traffic intensities.

Consider a matrix sequence $\mathscr{U} = \{U_i, 0 \leq i \leq n_t - 1\}$, where $U_i \in R^{n_s \times n_s}$, and define

$$f(\mathscr{U}) = \sum_{i=0}^{n_t-1} U_i \otimes ((\mathbf{t}\boldsymbol{\alpha})T^i). \tag{4.195}$$

It was shown in Alfa and Xue [3] that the matrix sequence $\{R(k),\ k \geq 0\}$ generated from the above iteration is of the form

$$R(k) = f(\mathscr{U}(k)), \tag{4.196}$$

where $\mathscr{U}(k) = \{U_i(k),\ 0 \leq i \leq n_t - 1\}$. If we define $\mathscr{V}(k) = \{V_i(k),\ 0 \leq i \leq n_t - 1\}$, where

$$V_0(k) = \mathbf{s}\boldsymbol{\beta}, \tag{4.197}$$

$$V_i(k) = V_{i-1}(k)S + U_{i-1}(k)\mathbf{s}\boldsymbol{\beta},\ 1 \leq i \leq n_t - 1, \tag{4.198}$$

then $\mathscr{U}(k)$ is determined from $\mathscr{V}(k)$ by

$$U_i(k+1) = S^{i+1} + S^*(I - V^*(k))^{-1} V_i(k),\ 0 \leq i \leq n_t - 1, \tag{4.199}$$

where

$$V^*(k) = \sum_{i=0}^{n_t-1} a_{i+1} V_i(k) \text{ and } S^* = \sum_{i=1}^{K} a_i S^i,$$

and

$$\tilde{K} = min(n_s - 1, n_t - 1).$$

In what follows we present the algorithm for computing the R matrix.

The Algorithm:

1. Set stopping tolerance ε
2. $S^* := a_1 S + a_2 S^2 + \cdot + a_{\tilde{K}} S^{\tilde{K}}$
3. $U_i^{new} := 0,\ i = 0, 1, 2, \cdots, n_t - 1$
4. Do
5. $U_i^{old} := U_i^{new},\ i = 0, 1, 2, \cdots, n_t - 1$
6. $V_0 := \mathbf{s}\boldsymbol{\beta}$
7. For $i = 1 : n_t - 1$
8. $V_i := V_{i-1} S + U_{i-1}^{old} \mathbf{s}\boldsymbol{\beta}$
9. End
10. $V^* := a_1 V_0 + a_2 V_1 + \cdots + a_{n_t} V_{n_t - 1}$
11. $V^* := S^*(I - V^*)^{-1}$
12. $V_0 := V^* V_0$
13. For $i = 1 : n_t - 1$
14. $V_i := V_{i-1} S + V^*(U_{i-1}^{old} \mathbf{s})\boldsymbol{\beta}$
15. End
16. $X_i^{new} := S^{i+1} + V_i,\ i = 0, 1, \cdots, n_t - 1$
17. Until $max_{i,j,k} |(U_i^{new})_{kl} - U_i^{old})_{kl}| < \varepsilon$.

4.10.3 The (ET,ET) representation for the GI/G/1 queue

Now we study the GI/G/1 system by considering the interarrival and service times as PH distributions using the elapsed time approach.. The arrival process is of PH distribution $(\boldsymbol{\alpha}, T)$ and service is PH distribution $(\boldsymbol{\beta}, S)$ written as

$$\boldsymbol{\alpha} = [1,\ 0,\ 0,\ \cdots,\ 0];\ \ \boldsymbol{\beta} = [1,\ 0,\ 0,\ \cdots,\ 0],$$

$$T_{i,j} = \begin{cases} \tilde{a}_i, & j = i+1 \\ 0, & otherwise, \end{cases},\quad S_{i,j} = \begin{cases} \tilde{b}_i, & j = i+1 \\ 0, & otherwise, \end{cases},$$

where

$$\tilde{a}_i = \frac{u_i}{u_{i-1}},\ u_i = 1 - \sum_{v=0}^{i} a_v,\ u_0 = 1,\ \tilde{a}_{n_t} = 0,$$

$$\tilde{b}_i = \frac{v_i}{v_{i-1}},\ v_i = 1 - \sum_{u=0}^{i} b_u,\ v_0 = 1,\ \tilde{b}_{n_s} = 0.$$

The DTMC representing the number in the system for this GI/G/1 queue is $\Delta = \{(0,k) \cup (i,k,j)\}; k = 1, 2, \cdots, n_t; j = 1, 2, \cdots, n_s; i \geq 1$, i.e. we have number in the system, phase of service and phase of arrival – in that order. The transition matrix for this system can be written as QBD as

$$P = \begin{bmatrix} B & C & & & \\ E & A_1 & A_0 & & \\ & A_2 & A_1 & A_0 & \\ & & \ddots & \ddots & \ddots \end{bmatrix},$$

with

$$B = T,\ C = \boldsymbol{\beta} \otimes (\mathbf{t}\boldsymbol{\alpha}),\ E = \mathbf{s} \otimes T,\ A_0 = S \otimes (\mathbf{t}\boldsymbol{\alpha}),$$
$$A_1 = (\mathbf{s}\boldsymbol{\beta}) \otimes (\mathbf{t}\boldsymbol{\alpha}) + S \otimes T,\ A_2 = (\mathbf{s}\boldsymbol{\beta}) \otimes T.$$

It is immediately clear that we can analyze this system using the matrix-geometric results as we did with the RT,RT case. However, the special feature of this model which gives us an additional structure which we can exploit is slightly different. We find that our block matrix A_2 has only one block of columns that is non-zero, as such the structure of our G matrix is

$$G = \begin{bmatrix} G_1 & 0 & 0 & \cdots & 0 \\ G_2 & 0 & 0 & \cdots & 0 \\ G_3 & 0 & 0 & \cdots & 0 \\ \vdots & \vdots & \vdots & \cdots & \vdots \\ G_{n_s} & 0 & 0 & \cdots & 0 \end{bmatrix}, \tag{4.200}$$

where G is the minimal non-negative solution to the matrix quadratic equation

$$G = A_2 + A_1 G + A_0 G^2. \tag{4.201}$$

Hence we try and compute the G matrix first using the block structure and then obtain the matrix R from it directly. By writing out the block matrix equations of G as follows

$$G_k = b_k^* T + b_k^* (\mathbf{t}\boldsymbol{\alpha}) G_k + \tilde{b}_k T G_{k+1} + \tilde{b}_k (\mathbf{t}\boldsymbol{\alpha}) G_{k+1} G_1,\ \ 1 \le k < n_s, \tag{4.202}$$

$$G_{n_s} = T + (\mathbf{t}\boldsymbol{\alpha}) G_1, \tag{4.203}$$

where $b_k^* = 1 - \tilde{b}_k$.

For simplicity we write these equations as

$$G_k = b_k^* V_0 + \tilde{b}_k V_0 G_{k+1} + b_k^* V_1 G_1 + \tilde{b}_k V_1 G_{k+1} G_1,\ \ 1 \le k < n_s, \tag{4.204}$$

$$G_{n_s} = b_{n_s}^* V_0 + b_{n_s}^* V_1 G_1, \tag{4.205}$$

keeping in mind that $b_{n_s}^* = 1$.

Let

$$\mathscr{H} = vecV_0 + (I \otimes V_1) vecG_1,\ \ \mathscr{J} = I \otimes V_0 + (G_1)^T \otimes V_1, \tag{4.206}$$

then using the same types of arguments as in the previous case we have

$$vecG_1 = [I - S(n_s-1,1)\mathscr{J}^{n_s-1}(I \otimes V_1)]^{-1}[S(n_s-1,1)\mathscr{J}^{n_s-1}vecV_0$$

$$+ b_1^*\mathscr{H} + \sum_{j=2}^{n_s-1} S(n_s-1,j)b_{n_s-j+1}^*\mathscr{J}^{n_s-1}\mathscr{H}] \qquad (4.207)$$

and

$$vecG_{n_s-k} = S(k,1)\mathscr{J}^k[vecV_0 + (I \otimes V_1)vecG_1] + b_{n_s-k}^*\mathscr{H}$$

$$+ \sum_{j=2}^{k} S(k,j)b_{n_s-j+1}^*\mathscr{J}^{k-j+1}\mathscr{H}, \ k = 1,2,\cdots,n_s-1, \qquad (4.208)$$

where $S(i,j) = \prod_{k=i}^{j} \tilde{b}_{n_s-i}$.

So all we need is to compute G_1 iteratively and then compute the remaining G_k, $k = 2,3,\cdots,n_s$ explicitly as above. A detail discussion of these results are available in Alfa [6].

4.11 GI/G/1/K Queues

In the following model, we represent the interarrival times as remaining time and interpret GI/G/1/K as a special case of M/G/1. For this end, at time $n(n \geq 0)$, let L_n be the remaining interarrival time, X_n be the number of customers in the system and J_n be the phase of the service time. Consider the state space

$$\Delta = \{(L_n, 0) \cup (L_n, X_n, J_n), n_t = 1,2,\cdots,n_t; X_n = 1,2,\cdots,K; J_n = 1,2,\cdots,n_s\}.$$

We also let $(L_n, X_n, J_n)|_{n\to\infty} = (L,X,J)$. In what follows, we refer to level $i (i \geq 1)$ as the set $\{((i,0) \cup (i,X,J), X = 1,2,\cdots,K; J = 1,2,\cdots,n_s\}$, and the second and third variables as the sub-levels.

This is indeed a DTMC. The transition matrix P_a representing the Markov chain for the GI/G/1/K system is

$$P_a = \begin{bmatrix} B_1 & B_2 & B_3 & \cdots & B_{n_t-1} & B_{n_t} \\ V & & & & & \\ & V & & & & \\ & & V & & & \\ & & & \ddots & & \\ & & & & V & 0 \end{bmatrix}, \qquad (4.209)$$

where

$$B_j = a_j \begin{bmatrix} 0 & \boldsymbol{\beta} & & & \\ & \mathbf{s}\boldsymbol{\beta} & S & & \\ & & \ddots & \ddots & \\ & & & \mathbf{s}\boldsymbol{\beta} & S \\ & & & & B \end{bmatrix}, \quad V = \begin{bmatrix} 1 & & & & \\ \mathbf{s} & S & & & \\ & \mathbf{s}\boldsymbol{\beta} & S & & \\ & & \ddots & \ddots & \\ & & & \mathbf{s}\boldsymbol{\beta} & S \end{bmatrix}$$

where $B = \mathbf{s}\boldsymbol{\beta} + S$ and $\mathbf{s} = \mathbf{1} - S\mathbf{1}$, and the matrices B_i, V are of dimensions $(Kn_s + 1) \times (Kn_s + 1)$.

The Markov chain is level-dependent and is a special case of the queue M/G/1. Let the stationary distribution of P_a be $\boldsymbol{x} = [\boldsymbol{x}_1, \boldsymbol{x}_2, \cdots, \boldsymbol{x}_{n_t}]$, where

$$\boldsymbol{x}_k = [x_{k,0}, \boldsymbol{x}_{k,1}, \cdots, \boldsymbol{x}_{k,K}] \text{ and } \boldsymbol{x}_{k,i} = [x_{k,i,1}, x_{k,i,2}, \cdots, x_{k,i,n_s}], i = 1, 2, \cdots, K.$$

The entry $x_{k,0}$ is the probability that the remaining interarrival time is k with no customers in the system; $x_{k,i,j}$ is the probability that the remaining interarrival time is k, the number of customers in the system is i and the phase of service of the customer in service is j.

Suppose that the Markov chain is positive recurrent, then the stationary vector $\boldsymbol{x}$ exists and satisfies

$$\boldsymbol{x}P_a = \boldsymbol{x}, \qquad \boldsymbol{x}\mathbf{1} = \mathbf{1}.$$

Writing $B_i = a_i U$, the equation $\boldsymbol{x}P_a = \boldsymbol{x}$ yields the recursion

$$\boldsymbol{x}_{n_t} = a_{n_t}\boldsymbol{x}_1 U \text{ and } \boldsymbol{x}_k = \boldsymbol{x}_{k+1}V + a_k\boldsymbol{x}_1 U, \quad k = n_t - 1, \cdots, 2, 1. \tag{4.210}$$

With this, we have the following result.

If the Markov chain is positive recurrent, we have

$$\boldsymbol{x}_k = \boldsymbol{x}_1 U \hat{a}_k(V), \; k = 1, 2, 3, \cdots, n_t, \tag{4.211}$$

where $\hat{a}_k(V) = \sum_{j=0}^{n_t-k} a_{k+j}V^j$, and $\boldsymbol{x}_1$ is normalized such that

$$\boldsymbol{x}_1\mathbf{1} = (\sum_{k=1}^{n_t} c_k)^{-1}, \qquad c_k = \sum_{i=k}^{n_t} a_i. \tag{3.4}$$

This is based on the following induction. For $k = n_t$, it is trivial. Suppose that Equation(4.211) is true for $k = s$, then for $k = s - 1$, we derive from 4.210 that

$$\boldsymbol{x}_{s-1} = \boldsymbol{x}_s V + a_{s-1}\boldsymbol{x}_1 U = \boldsymbol{x}_1 U \hat{a}_s(V)V + a_{s-1}\boldsymbol{x}_1 U = \boldsymbol{x}_1 U \hat{a}_{s-1}(V).$$

Thus from the induction process, Equation(4.211) is true for $k = 2, 3, \cdots, n_t$.

For evaluating $\boldsymbol{x}_1$, we substitute the formula for $\boldsymbol{x}_2, \cdots, \boldsymbol{x}_{n_t}$ into $\boldsymbol{x}P_a = \boldsymbol{x}$ and obtain

$$\boldsymbol{x}_1 = \boldsymbol{x}_1 B_1 + \boldsymbol{x}_2 V = \boldsymbol{x}_1 U[a_1 I + \hat{a}_2(V)V] = \boldsymbol{x}_1 U \hat{a}_1(V),$$

and the constraint $\boldsymbol{x}\mathbf{1} = 1$ yields

$$\boldsymbol{x}_1(I+c_2U+c_3UV+\cdots+c_{n_t}UV^{n_t-2})\mathbf{1}=1.$$

Noting that $c_1=1$ and U,V are stochastic, then

$$\boldsymbol{x}_1\mathbf{1}=(\sum_{k=1}^{n_t}c_k)^{-1}, \qquad c_k=\sum_{i=k}^{n_t}a_i.$$

Note that $\boldsymbol{x}_1$ here has similar structure as what Alfa presented in [17]. The reader can refer to [17] for the efficient computation of $\boldsymbol{x}_1$. In practical computations, we can use Equation (4.210) instead of Equation(4.211) to calculate $\boldsymbol{x}_k, k=2,\cdots,n_t$.

Further let $\mathbf{y}_k=\boldsymbol{x}_k\mathbf{1}$, then $\mathbf{y}=[\mathbf{y}_1,\cdots,\mathbf{y}_{n_t}]$ satisfies

$$\mathbf{y}=\mathbf{y}(T+\mathbf{t}\boldsymbol{\alpha}) \text{ and } \mathbf{y}\mathbf{1}=1,$$

where $(\boldsymbol{\alpha},T)=(\boldsymbol{\alpha}_r,T_r)$. This is a good criteria for checking whether $\boldsymbol{x}$ is correctly calculated.

4.11.1 Queue length

Recall the variable $x_{k,0}$ is the probability that the remaining interarrival time is k with no customers in the system; $x_{k,i,j}$ is the probability that the remaining interarrival time is k, the number of customers in the system is i and the phase of service of the customer in service is j. Hence, if we let q_i be the probability of having i customers in the system. Then

$$q_0=\sum_{k=1}^{n_t}x_{k,0}=\sum_{k=1}^{n_t}\boldsymbol{x}_k\mathbf{e}_1, \tag{4.212}$$

$$q_i=\sum_{k=1}^{n_t}\sum_{j=1}^{n_s}x_{k,i,j}=\sum_{k=1}^{n_t}\boldsymbol{x}_k\hat{\mathbf{e}}_i,\ i=1,2,\cdots,K, \tag{4.213}$$

where $\hat{\mathbf{e}}_i=\sum_{j=2+(i-1)n_s}^{1+in_s}\mathbf{e}_j$.

4.11.2 Waiting times

In order to study the waiting time distribution, we need to obtain the state of the system just after the arrival of an arbitrary customer. Let $\mathscr{U}$ be the number of customers in the system as seen by an arriving and unblocked customer at the steady state. Let

$$\psi_{k,0} = Pr\{L = k, \mathscr{U} = 0\}, \qquad \psi_{k,i,j} = Pr\{L = k, \mathscr{U} = i, J = j\}$$

Further let

$$\boldsymbol{\psi}_0 = [\psi_{1,0}, \psi_{2,0}, \cdots, \psi_{n_t,0}], \boldsymbol{\psi}_i = [\boldsymbol{\psi}_{1,i}, \boldsymbol{\psi}_{2,i}, \cdots, \boldsymbol{\psi}_{n_t,i}],$$

and

$$\boldsymbol{\psi}_{k,i} = [\psi_{k,i,1}, \psi_{k,i,2}, \cdots, \psi_{k,i,n_s}],$$

then by conditioning on the state of the observation of the system at arrivals, we obtain

$$\boldsymbol{\psi}_0 = h^{-1}[\boldsymbol{x}_0(\mathbf{t}\boldsymbol{\alpha}) + \boldsymbol{x}_1(\mathbf{t}\boldsymbol{\alpha}) \otimes \mathbf{s}], \tag{4.214}$$

$$\boldsymbol{\psi}_i = h^{-1}[\boldsymbol{x}_i((\mathbf{t}\boldsymbol{\alpha}) \otimes S) + \sum_{i=1}^{K-1} \boldsymbol{x}_{i+1}(\mathbf{t}\boldsymbol{\alpha}) \otimes \mathbf{s}\boldsymbol{\beta}], \tag{4.215}$$

where $\boldsymbol{x}_0 = [x_{1,0}, x_{2,0} \cdots, x_{n_t,0}], \boldsymbol{x}_i = [\boldsymbol{x}_{1,i}, \boldsymbol{x}_{2,i}, \cdots, \boldsymbol{x}_{n_t,i}]$, and

$$h = [\boldsymbol{x}_0(\mathbf{t}\boldsymbol{\alpha}) + \boldsymbol{x}_1(\mathbf{t}\boldsymbol{\alpha}) \otimes \mathbf{s}]\mathbf{1} + [\boldsymbol{x}_i((\mathbf{t}\boldsymbol{\alpha}) \otimes S) + \sum_{i=1}^{K-1} \boldsymbol{x}_{i+1}(\mathbf{t}\boldsymbol{\alpha}) \otimes (\mathbf{s}\boldsymbol{\beta})]\mathbf{1}.$$

Let $\mathscr{W}$ be the waiting time of a customer. Define

$$W_r = Pr\{\mathscr{W} \leq r\}, \qquad w_r = Pr\{\mathscr{W} = r\}.$$

The waiting time of an arbitrary customer is the workload in the system as seen by him, at his arrival. Hence, we have

$$w_0 = \boldsymbol{\psi}_0 \mathbf{1}, \tag{4.216}$$

$$w_r = \sum_{i=1}^{r} \boldsymbol{\psi}_i (\mathbf{1}_{n_t} \otimes I_{n_s}) B_r^i \mathbf{1}_{n_s}, \quad r \geq 1, \tag{4.217}$$

where B_r^i satisfy

$$B_0^0 = I_{n_s}, \quad B_r^r = (\mathbf{s}\boldsymbol{\beta})^r, \quad B_r^0 = 0, \quad B_r^1 = S^{r-1}(\mathbf{s}\boldsymbol{\beta}), \; r \geq 1,$$
$$B_r^i = (\mathbf{s}\boldsymbol{\beta})B_{r-1}^{i-1} + SB_{r-1}^i, \; 2 \leq i \leq r.$$

where the matrix B_r^i represents the probability distribution of the work in the system at the arrival of a customer who find i customers ahead of him, and $(B_r^i)_{u,v}$ represents the probability that the phase of the service of the customer in service at his arrival of the arbitrary customer is u and at the completion of the i customer service it is in phase v, and that the workload associated with the i customers in the system is r units of time.

To derive the distribution of W_r, we need to define $\hat{X}_n$ be the number of customers ahead of the arbitrary customer in the queue, and let $\hat{J}_n$ be the service phase of the customer in service. Then the state space of the absorbing Markov chain is given as

$$\Delta^w = \{(0) \cup (\hat{X}_n, \hat{J}_n), \hat{X}_n = 1,2,\cdots,K-1; \hat{J}_n = 1,2,\cdots,n_s\},$$

and the associated transition matrix is just the $((K-1)n_s+1) \times ((K-1)n_s+1)$ principle submatrix of V. We denote it as V_w. Let $\mathbf{z}^{(0)} = [z_0^{(0)}, \mathbf{z}_1^{(0)}, \mathbf{z}_2^{(0)}, \cdots, \mathbf{z}_{K-1}^{(0)}]$, where $z_0^{(0)} = \Psi_0 \mathbf{1}_{n_t}, \mathbf{z}_i^{(0)} = \psi_i^{(0)}(\mathbf{1}_{n_t} \otimes I_{n_s})$, then we have

$$\mathbf{z}^{(i)} = \mathbf{z}^{(i-1)} V_w = \mathbf{z}^{(0)} V_w^i, \quad i \geq 1,$$

and

$$W_r = z_0^{(r)}.$$

4.11.3 Model II

In this section, we consider the GI/G/1/K system with the interarrival time with representation (α, T) and service times with representation (β, S). We use the remaining service time as the first variable, i.e. the level, for the state space of GI/G/1/K system. Let L_n be the phase of the arrival, X_n be the number of the customers in the system, J_n be the remaining service time of the customer in service at time n. Consider the state space

$$\Delta = \{(0, L_n) \cup (J_n, X_n, L_n), J_n = 1,2,\cdots,n_s; X_n = 1,\cdots,K; L_n = 1,\cdots,n_t\}.$$

It is obvious that Δ is indeed a Markov chain with transition matrix P_{sr} as

$$P_{sr} = \begin{bmatrix} T & E & & & & & \\ D_0 & D_1 & D_2 & D_3 & \cdots & D_{n_s-1} & D_{n_s} \\ & H & & & & & \\ & & H & & & & \\ & & & H & & & \\ & & & & \ddots & & \\ & & & & & H & 0 \end{bmatrix},$$

where $E = [\mathbf{t}\alpha, \hat{\mathbf{0}}]$, $D_0 = \begin{bmatrix} T \\ \hat{\mathbf{0}}' \end{bmatrix}$ with $\hat{\mathbf{0}}$ of dimension $n_t \times (K-1)n_t$, and for $j \geq 1$,

$$D_j = b_j \begin{bmatrix} \mathbf{t}\alpha & & & \\ T & \mathbf{t}\alpha & & \\ & \ddots & \ddots & \\ & & T & \mathbf{t}\alpha \end{bmatrix}, \quad H = \begin{bmatrix} T & \mathbf{t}\alpha & & \\ & \ddots & \ddots & \\ & & T & \mathbf{t}\alpha \\ & & & D \end{bmatrix},$$

where $D = \mathbf{t}\alpha + T$ and $\mathbf{t} = \mathbf{1} - T\mathbf{1}$, and the matrices $D_i (i \geq 1), H$ are of dimensions $(Kn_t) \times (Kn_t)$.

Obviously, P_{sr} is also a special case of M/G/1, and it is easy to show that P_{sr} is positive recurrent always. Hence the associated stationary distribution exists. Let the stationary distribution of P_{sr} be $\mathbf{y} = [\mathbf{y}_0, \mathbf{y}_1, \mathbf{y}_2, \cdots, \mathbf{y}_{n_s}]$, where

$$\mathbf{y}_0 = [y_{0,1}, y_{0,2}, \cdots, y_{0,n_t}], \quad \mathbf{y}_k = [\mathbf{y}_{k,1}, \cdots, \mathbf{y}_{k,K}],$$

and $\mathbf{y}_{k,i} = [y_{k,i,1}, y_{k,i,2}, \cdots, y_{k,i,n_t}], i = 1, 2, \cdots, K$.

Writing $D_j = b_j F$, $j = 1, 2, \cdots, m$, we have

$$\mathbf{y}_k = \mathbf{y}_1 F \hat{b}_k(H), \quad k = 2, 3, \cdots, n_s, \tag{4.218}$$

where $\hat{b}_k(H) = \sum_{j=0}^{n_s - k} b_{k+j} H^j$, and $[\mathbf{y}_0, \mathbf{y}_1]$ is the left eigenvector which corresponds to the eigenvalue of 1 for the stochastic matrix

$$\begin{bmatrix} T & E \\ D_0 & D_{11} \end{bmatrix},$$

where $D_{11} = \sum_{j=1}^{n_s} b_j F H^{j-1}$ records the probability, starting from level 1, of reaching level 1 before level 0, and $[\mathbf{y}_0, \mathbf{y}_1]$ is normalized with

$$\mathbf{y}_0 \mathbf{1} + \mathbf{y}_1 [I + (\sum_{k=2}^{n_s} \hat{c}_k) F] \mathbf{1} = 1, \tag{4.219}$$

where $\hat{c}_k = \sum_{v=k}^{n_s} b_v$.

The arguments leading to this result are as follows. The equality $\mathbf{y} P_{sr} = \mathbf{y}$ yields

$$\mathbf{y}_{n_s} = \mathbf{y}_1 D_{n_s} \quad \text{and} \quad \mathbf{y}_k = \mathbf{y}_1 D_k + \mathbf{y}_{k+1} H, \quad k = 2, 3, \cdots, n_s.$$

From this recursion and induction process, it is easy to show that Equation (4.218) is true for $k = 2, 3, \cdots, n_s$.

Also from the formula Equation (4.218) for $\mathbf{y}_2$ and the equation

$$[\mathbf{y}_0 \quad \mathbf{y}_1] = [\mathbf{y}_0 \quad \mathbf{y}_1 \quad \mathbf{y}_2] \begin{bmatrix} T & E \\ D_0 & D_1 \\ 0 & H \end{bmatrix} = [\mathbf{y}_0 \quad \mathbf{y}_1] \begin{bmatrix} T & E \\ D_0 & D_{11} \end{bmatrix},$$

we obtain

$$D_{11} = D_1 + F\hat{b}_2(H)H = D_1 + D_2H + \cdots + D_{n_s}H^{n_s-1} = \sum_{j=1}^{n_s} b_jFH^{j-1}.$$

Now the equation $\mathbf{y1} = 1$ leads to

$$\mathbf{y}_0\mathbf{1} + \mathbf{y}_1\mathbf{1} + \sum_{k=2}^{n_s} \mathbf{y}_1F\hat{b}_k(H)\mathbf{1} = 1.$$

We consider the same GI/G/1/K system but with J_n representing the elapsed service time of the customer in service at time n, L_n be the phase of the arrival, X_n be the number of the customers in the system. We present only Model II and the corresponding transition matrix associated with the DTMC. The transition matrix P_{se} should be

$$P_{se} = \begin{bmatrix} T & E & & & & \\ C_1 & B_1 & A_1 & & & \\ C_2 & B_2 & & A_2 & & \\ \vdots & \vdots & & & \ddots & \\ C_{n_s-1} & B_{n_s-1} & & & & A_{n_s-1} \\ C_{n_s} & B_{n_s} & & & & 0 \end{bmatrix},$$

where $B_j = (1-\tilde{b}_j)F$, $A_j = \tilde{b}_jH$, $C_j = (1-\tilde{b}_j)D_0$, and F, H, D_0, E are defined in Section 4.11. Let the stationary distribution of P_{se} be $\mathbf{y} = [\mathbf{y}_0, \mathbf{y}_1, \mathbf{y}_2, \cdots, \mathbf{y}_{n_s}]$, where $\mathbf{y}_i$ are defined earlier. Similarly, we have the following result.

For this we have

$$\mathbf{y}_k = \mathbf{y}_{k-1}A_{k-1},\ k = 2,3,\cdots,n_s, \tag{4.220}$$

and $[\mathbf{y}_0, \mathbf{y}_1]$ is the left eigenvector which corresponds to the eigenvalue of 1 for the stochastic matrix

$$\begin{bmatrix} T & E \\ \hat{b}_1(H)D_0 & \hat{b}_1(H)F \end{bmatrix},$$

where $\hat{b}_1(H) = \sum_{j=1}^{n_s-1} b_jH^{j-1}$ and $[\mathbf{y}_0, \mathbf{y}_1]$ is normalized such that

$$\mathbf{y}_0\mathbf{1} + (\sum_{k=1}^{n_s} \hat{c}_k)\mathbf{y}_1\mathbf{1} = 1,$$

where $\hat{c}_k = \sum_{v=k}^{n_s} b_v$.

Note that the most computationally involved step in Model I-II is evaluating the matrices $\hat{a}_1(V)$ and $\hat{b}_1(H)$, respectively. Therefore, Model I is more appropriate when $n_t > n_s$, and Model II is good when $n_t < n_s$. When $n_t = n_s$, either method is appropriate.

The idea of re-arranged state space is easily extended to the $GI^X/G/1/K$ system. For details see Liu, Alfa and Xue [70].

4.12 MAP/PH/1 Queues

The MAP/PH/1 queue is a generalization of the PH/PH/1 queue. If we let the arrivals be described by the two sub-stochastic matrices D_0 and D_1 of order n_t, with $D = D_0 + D_1$ being a stochastic matrix. Letting the service times have a phase type distribution with representation $(\boldsymbol{\beta}, S)$ of dimension n_s, then we can simply analyze this system by replacing the block matrices of the transition matrix of the PH/PH/1 queue as follows. Wherever T appears in the PH/PH/1 results, it should be replaced with D_0 and wherever $\mathbf{t}^0\alpha$ appears in the PH/PH/1 queue it should be replaced with D_1. Consider the transition probability matrix of the PH/PH/1 system written as

$$P = \begin{bmatrix} B & C & & & \\ E & A_1 & A_0 & & \\ & A_2 & A_1 & A_0 & \\ & & \ddots & \ddots & \ddots \end{bmatrix},$$

then the block matrices for this MAP/PH/1 has the following expressions

$$B = D_0,\ C = D_1 \otimes \boldsymbol{\beta},\ A_0 = D_1 \otimes S,\ A_1 = D_0 \otimes S + D_1 \otimes (\mathbf{s}\boldsymbol{\beta}),\ A_2 = D_0 \otimes (\mathbf{s}\boldsymbol{\beta}).$$

The rest of the analysis is left to the reader as an excercise.

4.13 Batch Queues

Batch queues are systems in which at each arrival epoch the number of items that arrive could be more than one. This is very common in telecommunication systems where packets arrive in batches. Also we may have batch services where more than one item are served together, such as public transportation system where customers are served in groups depending on the number waiting, the server capacity and protocols. This also happens in telecommunications where messages arrive in batches; for example in the ATM system of communication, packets of different sizes arrive at the system and then the packets re-grouped into several cells of equal sizes – these cells will be the batch arrivals in this case.

4.13.1 Geo^X/Geo/1 queue

Usually because we are working in discrete time and the size of the time interval, otherwise known as the time slot in telecommunications, affects how we account for the arrival. There is a high chance that if there is an arrival in a time slot that arrival could be more than one. Hence considering a queueing system with batch arrivals is very important in telecommunications.

Here we consider a case where arrival is Bernoulli with $a = \Pr\{$an arrival in a time slot$\}$ and $\bar{a}$ its complement. However when arrivals occur they could be in batch. So we define $\theta_i = \Pr\{$number of arrivals in a time slot is $i\}$, given there is an arrival, with $i \geq 1$. We let $\bar{\theta} = \sum_{i=1}^{\infty} i\theta$. Let us further write $c_0 = \bar{a}$ and $c_i = a\theta_i,\ i \geq 1$. We therefore have a vector $\mathbf{c} = [c_0,\ c_1,\ \cdots]$ which represents arrivals in a time slot. The mean arrival rate is thus given as $\lambda = a\bar{\theta}$. We also let the service be geometric with parameter b. Consider the DTMC $\{X_n,\ n \geq 0\}$ with state space $\{0,1,2,\cdots\}$, where X_n is the number of items in the system at time n. It is straightforward to show that the transition matrix of this DTMC is

$$P = \begin{bmatrix} c_0 & c_1 & c_2 & c_3 & \cdots \\ a_0 & a_1 & a_2 & a_3 & \cdots \\ & a_0 & a_1 & a_2 & \cdots \\ & & a_0 & a_1 & \cdots \\ & & & \ddots & \ddots \end{bmatrix}, \tag{4.221}$$

where c_i are as defined above and

$$a_0 = bc_0,\ a_i = bc_i + \bar{b}c_{i-1},\ i \geq 1.$$

This is a DTMC of the M/G/1 type and all the matrix-analytic results for the M/G/1 apply. Provided the system is stable, i.e. provided $\lambda < b$, then we have a unique vector $\boldsymbol{x} = [x_0,\ x_1,\ x_2,\ \cdots]$ such that

$$\boldsymbol{x} = \boldsymbol{x}P,\ \ \boldsymbol{x}\mathbf{1} = 1.$$

We have the general relationship

$$x_i = x_0 c_i + \sum_{j=1}^{i+1} x_j a_{i-j+1},\ \ i \geq 0. \tag{4.222}$$

We can apply $z-$transform approach or the matrix-analytic approach to analyze this problem.

4.13.1.1 Transform Approach

If we define

$$X^*(z) = \sum_{i=0}^{\infty} z^i x_i,\ \ c^*(z) = \sum_{i=0}^{\infty} z^i c_i,\ \text{and}\ \ a^*(z) = \sum_{i=0}^{\infty} z^i a_i,\ \ |z| \leq 1,$$

then we have

$$X^*(z) = \frac{x_0[zc^*(z) - a^*(z)]}{z - a^*(z)}. \tag{4.223}$$

We know that $X^*(z)|_{z\to 1} = 1$. However, we have to apply l'Hôpital's rule here. First we show that

$$zc^*(z) = z(\bar{a} + a\theta^*(z)), \tag{4.224}$$

hence

$$\frac{d(zc^*(z))}{dz}|_{z\to 1} = 1 + a\bar{\theta},$$

implying that $\bar{c} = \sum_{i=1}^{\infty} ic_i = a\bar{\theta}$. We can also show that

$$\frac{d(a^*(z))}{dz}|_{z\to 1} = b\bar{c} + \bar{b}\bar{c} + \bar{b} = a\bar{\theta} + \bar{b}. \tag{4.225}$$

Now we have

$$X^*(z)|_{z\to 1} = 1 = x_0[1 + a\bar{\theta} - \bar{b} - a\bar{\theta}][1 - \bar{b} - a\bar{\theta}]^{-1} = x_0 b[1 - a\bar{\theta} - \bar{b}]^{-1}, \tag{4.226}$$

which gives

$$x_0 = \frac{b - a\bar{\theta}}{b} = 1 - \frac{a\bar{\theta}}{b}. \tag{4.227}$$

Hence we write

$$X^*(z) = \frac{(b - a\bar{\theta})[zc^*(z) - a^*(z)]}{b(z - a^*(z))}. \tag{4.228}$$

4.13.1.2 Examples

- **Example 1: Uniform batch size:** We consider the case in which the batch size varies from k_1 to k_2 with θ_i being the probability that the arriving batch size is i with $\theta_i = 0,\ i < k_1; i > k_2$ and both $(k_1, k_2) < \infty$.
  For this example, $\theta^*(z) = \sum_{j=k_1}^{k_2} z^i \theta_i$ and $\bar{\theta} = \sum_{j=k_1}^{k_2} j\theta_j$. For example, when $k_1 = k_2 = 3$ we have $\theta^*(z) = z^3$ and $\bar{\theta} = 3$. For this we have

$$X^*(z) = \frac{(b - 3a)[z^4 - a^*(z)]}{b(z - a^*(z))}.$$

- **Example 2: Geometric batch size:** We consider the case in which the batch size has the geometric distribution with parameter θ and $\theta_i = \theta(1-\theta)^{i-1},\ i \geq 1$. In this case $\theta^*(z) = z(1-\theta)(1 - z\theta)^{-1}$ and $\bar{\theta} = \theta^{-1}$. For this example we have

$$X^*(z) = \frac{(b - \frac{a}{\theta})[z^2(1-\theta) - a^*(z)(1 - z\theta]}{b(z - a^*(z))}.$$

  In both examples we simply substitute $\theta^*(z)$ appropriately to the general transform results presented earlier, Equation (4.229), and then obtain $X^*(z)$.

4.13.1.3 Matrix-analytic Approach

By virtue of the fact that if this system is stable, then the G matrix which is a scalar here and we call it g, will be unity. In this case we can easily apply the recursion by Ramaswami [89] as follows.

For any size $k \geq 1$ we write the transition matrix

$$P(k) = \begin{bmatrix} c_0 & c_1 & c_2 & \cdots & c_{k-1} & \sum_{j=k}^{\infty} c_j \\ a_0 & a_1 & a_2 & \cdots & a_{k-1} & \sum_{j=k}^{\infty} a_j \\ & a_0 & a_1 & \cdots & a_{k-2} & \sum_{j=k-1}^{\infty} a_j \\ & & a_0 & \cdots & a_{k-3} & \sum_{j=k-2}^{\infty} a_j \\ & & & \ddots & \ddots & \vdots \\ & & & & a_0 & \sum_{j=1}^{\infty} a_j \end{bmatrix}, \quad (4.229)$$

which can be written as

$$P(k) = \begin{bmatrix} c_0 & c_1 & c_2 & \cdots & c_{k-1} & \tilde{c}_k \\ a_0 & a_1 & a_2 & \cdots & a_{k-1} & \tilde{a}_k \\ & a_0 & a_1 & \cdots & a_{k-2} & \tilde{a}_{k-1} \\ & & a_0 & \cdots & a_{k-3} & \tilde{a}_{k-2} \\ & & & \ddots & \ddots & \vdots \\ & & & & a_0 & \tilde{a}_1 \end{bmatrix}, \quad (4.230)$$

where $\tilde{c}_k = \sum_{j=k}^{\infty} c_j, \ \tilde{a}_{k-v} = \sum_{j=k-v}^{\infty} a_j$.

Hence we can write

$$x_k = (1-\tilde{a}_1)^{-1}[x_0\tilde{c}_k + \sum_{j=1}^{k-1} x_j \tilde{a}_{k-j+1}], \ k = 1,2,\cdots. \quad (4.231)$$

Keeping in mind that the x_0 is known then the recursion starts from $k = 2$ and continues to the point of truncation.

4.13.2 Geo/Geo^Y/1 queue

In some systems packets are served in batches, rather than individually. Here we consider a case where arrival is Bernoulli with $a = \Pr\{\text{an arrival in a time slot}\}$ and $\bar{a}$ its complement. Service is geometric with parameter b and also in batches of size $\mathscr{B}$, with $\theta_j = Pr\{\mathscr{B} = i, \ i \geq 1\}$, with $\sum_{j=1}^{\infty} \theta_j = 1$. We can write $b_0 = 1 - b$ and $b_i = b\theta_i$. We therefore have a vector $\mathbf{b} = [b_0, \ b_1, \ \cdots]$ which represents the number of packets served in a time slot. The mean service rate is thus given as $\mu = b\bar{\theta}$, where $\bar{\theta} = \sum_{j=1}^{\infty} j\theta_j$. Consider the DTMC $\{X_n, \ n \geq 0\}$ with state space $\{0,1,2,\cdots\}$, where X_n is the number of items in the system at time n. It is straightforward to show that the transition matrix of this DTMC is

$$P = \begin{bmatrix} c_0 & d_1 & & & & \\ c_1 & d_1 & d_0 & & & \\ c_2 & d_2 & d_1 & d_0 & & \\ c_3 & d_3 & d_2 & d_1 & d_0 & \\ \vdots & \vdots & \vdots & \vdots & \vdots & \ddots \end{bmatrix},$$

where

$$c_0 = 1-a,\ d_0 = a,\ d_j = ab_j + (1-a)b_{j-1},\ j \geq 1,\ c_j = 1 - \sum_{k=0}^{j} d_k,\ j \geq 1.$$

It is immediately clear that we have the GI/M/1 type DTMC and we can apply the related results for analyzing this problem. First we find the variable r which is the minimal non-negative solution to the non-linear equation

$$r = \sum_{j=0}^{\infty} a_j r^j.$$

If the system is stable, i.e. $a < b\bar{\theta}$ then we have a solution

$$\boldsymbol{x} = \boldsymbol{x}P,\ \ \boldsymbol{x}\mathbf{1} = 1,$$

where $\boldsymbol{x} = [x_0,\ x_1,\ x_2,\ \cdots]$, and

$$x_{i+1} = x_i r.$$

The boundary equation is solved as

$$x_0 = \sum_{j=0}^{\infty} x_j c_j = x_0 \sum_{j=0}^{\infty} r^j c_j.$$

This is then normalized by $x_0(1-r)^{-1} = 1$.

Chapter 5
Advance Single Node Queuing Models

5.1 Multiserver Queues

Multiserver queueing systems are very useful in modelling telecommunication systems. Usually in such systems we have several channels that are used for communications. These are considered as parallel servers in a queueing system. Throughout this chapter we will be dealing with cases where the parallel servers are identical. The case of heterogeneous servers will not be covered in this book.

Two classes of multiserver queues will be dealt with in detail; the Geo/Geo/k and the PH/PH/k. We will show later that most of the other multiserver queues to be considered are either special cases of the PH/PH/k or general cases of the Geo/Geo/k systems.

One key difference between discrete time and continuous time analyses of queueing systems comes when studying multiserver systems. As long as a single server queue has single arrivals and only one item is served at a time by each server then for most basic continuous time models the system is BD such as the M/M/s or QBD such as the M/PH/s. This is not the case with the discrete time systems. As will be seen in this section, the simple Geo/Geo/k system is not a BD. It is a special case of the GI/M/1 type but may be converted to a QBD. This key difference comes because it is possible to have more than one job completion at any time, when there is more than one job receiving service. This difference makes the discrete time multiserver queues more involved in terms of analysis.

5.1.1 Geo/Geo/k Systems

In the Geo/Geo/k queue we have packets arriving according to the Bernoulli process, i.e. the inter-arrival times have geometric distribution with parameter $a = Pr\{an\ arrival\ at\ any\ time\ epoch\}$, with $\bar{a} = 1 - a$. There are $k < \infty$ identical servers in parallel each with a geometric service distribution with parameter b and $\bar{b} = 1 - b$.

A.S. Alfa, *Queueing Theory for Telecommunications*,
DOI 10.1007/978-1-4419-7314-6_5, © Springer Science+Business Media, LLC 2010

Since the severs are all identical and have geometric service process, we know that if there are j packets in service, then the probability of i $(i \leq j)$ of those packets having their service completed in one time slot, written as b_i^j is given by a binomial distribution, i.e.

$$b_i^j = \binom{j}{i} b^i (1-b)^{j-i}, \; 0 \leq i \leq j. \tag{5.1}$$

Consider the DTMC $\{X_n, \, n \geq 0\}$ which has the state space $\{0,1,2,3,\cdots\}$ where X_n is the number of packets in the system at time slot n. If we define $p_{i,j} = Pr\{X_{n+1} = j | X_n = i\}$, then we have

$$p_{0,0} = \bar{a}, \tag{5.2}$$

$$p_{0,1} = a, \tag{5.3}$$

$$p_{0,j} = 0, \; \forall j \geq 2, \tag{5.4}$$

and

$$p_{i,j} = \begin{cases} ab_0^i, & j = i+1, \; i < k \\ \bar{a}b_i^i, & j = 0, \quad i < k \\ ab_{i-j+1}^i + \bar{a}b_{i-j}^i, & 1 \leq j \leq i \\ 0, & j > i+1, \; i < k \end{cases}, \tag{5.5}$$

and

$$p_{i,j} = \begin{cases} c_0 = ab_0^k, & j = i+1, \quad i \geq k \\ c_{k+1} = \bar{a}b_k^k, & j = i-k, \quad i \geq k \\ c_{i-j+1} = ab_{i-j+1}^k + \bar{a}b_{i-j}^k, & 1 \leq j \leq i, \; i \geq k, \\ 0, & j > i+1, \quad i \geq k \\ 0, & i > k, \quad j \leq i-k \end{cases} \tag{5.6}$$

It is immediately clear that the DTMC is of the GI/M/1 type, with $p_{i,j} = 0, \; i > k, \; j \leq i-k$. Hence we can easily use the results from the GI/M/1 to analyze this system.

5.1.1.1 As GI/M/1 type

We display the case of $k = 3$ first and then show the general case in a block matrix form. For $k = 3$ the transition matrix of this DTMC is given as

$$P = \begin{bmatrix} \bar{a} & a & & & & & \\ \bar{a}b_1^1 & ab_1^1 + \bar{a}b_0^1 & ab_0^1 & & & & \\ \bar{a}b_2^2 & ab_2^2 + \bar{a}b_1^2 & ab_1^2 + \bar{a}b_0^2 & ab_0^2 & & & \\ \bar{a}b_3^3 & ab_3^3 + \bar{a}b_2^3 & ab_2^3 + \bar{a}b_1^3 & ab_1^3 + \bar{a}b_0^3 & ab_0^3 & & \\ & \bar{a}b_3^3 & ab_3^3 + \bar{a}b_2^3 & ab_2^3 + \bar{a}b_1^3 & ab_1^3 + \bar{a}b_0^3 & ab_0^3 & \\ & & \bar{a}b_3^3 & ab_3^3 + \bar{a}b_2^3 & ab_2^3 + \bar{a}b_1^3 & ab_1^3 + \bar{a}b_0^3 & ab_0^3 \\ & & & \ddots & \ddots & \ddots & \ddots \; \ddots \end{bmatrix},$$

which we write as

$$P = \begin{bmatrix} p_{0,0} & p_{0,1} & & & & & & \\ p_{1,0} & p_{1,1} & p_{1,2} & & & & & \\ p_{2,0} & p_{2,1} & p_{2,2} & p_{2,3} & & & & \\ c_4 & c_3 & c_2 & c_1 & c_0 & & & \\ & c_4 & c_3 & c_2 & c_1 & c_0 & & \\ & & c_4 & c_3 & c_2 & c_1 & c_0 & \\ & & & \ddots & \ddots & \ddots & \ddots & \ddots \end{bmatrix}. \quad (5.7)$$

We consider a stable system. Our interest is to solve for $\boldsymbol{x} = \boldsymbol{x}P,\ \boldsymbol{x}\mathbf{1} = 1$, where $\boldsymbol{x} = [x_0,\ x_1,\ x_2,\ \cdots]$. We first present the GI/M/1 approach for this problem.

It is clear that the rows of the matrix P are repeating after the $(k+1)^{st}$ row. Let the repeating part be written as $[c_{k+1},\ c_k,\ \cdots,\ c_0]$. It is known from that this DTMC is stable if

$$\beta > 1,$$

where $\beta = \sum_{v=0}^{k+1} v c_v$ which gives $\beta = kb + \bar{a} > 1$. Hence provided that

$$kb > a \quad (5.8)$$

then the system is stable. For a stable system we have a scalar r which is the minimum non-negative solution to the equation

$$r = \sum_{j=0}^{k+1} r^j c_j. \quad (5.9)$$

It is easy to see that r is the solution to the polynomial equation

$$r = (br + \bar{b})^k (a + \bar{a}r). \quad (5.10)$$

Using the GI/M/1 results we have

$$x_{i+k} = x_k r^i,\ i \geq 0. \quad (5.11)$$

However, we still need to solve for the boundary values, i.e. $[x_0,\ x_1,\ \cdots,\ x_k]$. Solving the boundary equations reduces to a finite DTMC problem. For the case of $k = 3$ we display the transition matrix corresponding to this boundary behaviour, using $p_{i,j}$ as elements of this finite DTMC. We call the transition matrix P_b and write it as follows:

$$P_b = \begin{bmatrix} p_{0,0} & p_{0,1} & & \\ p_{1,0} & p_{1,1} & p_{1,2} & \\ p_{2,0} & p_{2,1} & p_{2,2} & p_{2,3} \\ c_4 & \sum_{i=3}^{4} r^{i-3} c_i & \sum_{i=2}^{4} r^{i-2} c_i & \sum_{i=1}^{4} r^{i-1} c_i \end{bmatrix}. \quad (5.12)$$

Now going back to the more general case from here we can write this equation as

$$P_b = \begin{bmatrix} q_{0,0} & q_{0,1} & & & & \\ q_{1,0} & q_{1,1} & q_{1,2} & & & \\ q_{2,0} & q_{2,1} & q_{2,2} & q_{2,3} & & \\ \vdots & \vdots & \vdots & \vdots & \ddots & \\ q_{k,0} & q_{k,1} & q_{k,2} & q_{k,3} & \cdots & q_{k,k+1} \\ q_{k+1,0} & q_{k+1,1} & q_{k+1,2} & q_{k+1,3} & \cdots & q_{k+1,k+1} \end{bmatrix}, \tag{5.13}$$

and solve for the boundary values as

$$[x_0\ x_1\ x_2\ \cdots\ x_k] = [x_0\ x_1\ x_2\ \cdots\ x_k]P_b. \tag{5.14}$$

This can be solved easily starting from

$$x_k = \frac{x_{k+1}(1-q_{k+1,k+1})}{q_{k,k+1}}, \tag{5.15}$$

and carrying out a backward substitution recursively from $j = k-1$ up to $j = 1$ and into

$$x_j = \sum_{v=j-1}^{k+1} x_v q_{v,j}, \quad 1 \le j \le k-1, \tag{5.16}$$

$$x_0 = \sum_{v=0}^{k+1} x_v q_{v,0}. \tag{5.17}$$

This is then normalized as

$$\sum_{j=0}^{\infty} x_j = 1, \quad \rightarrow x_k(1-r)^{-1} + \sum_{j=0}^{k-1} x_j = 1. \tag{5.18}$$

The distribution of $X_n|_{n\to\infty}$ is now fully determined, using the standard method.

We point out that Artalejo and Hernandez-Lerma [19] did discuss the above-mentioned recursion for obtaining the boundary values.

Next we consider studying this problem as a QBD DTMC.

5.1.1.2 As a QBD

There are more results for QBDs than any other DTMC with block structures in the literature. So whenever it is possible to convert a problem to the QBD structure it is usually an advantage. We study the Geo/Geo/k problem as a QBD. Consider a bivariate DTMC $\{(Y_n, J_n), n \ge 0\}$, $Y_n = 0, 1, 2, \cdots$, $J_n = 0, 1, 2, \cdots, k-1$. At anytime n we let $kY_n + J_n$ be the number of packets in the system, i.e. $kY_n + J_n = X_n$. For example, for the case of $k = 3$ the situation with $Y_n = 5, J_n = 2$ implies that we have $3 \times 5 + 2 = 17$ packets in the system, including the ones receiving service. What exactly we have done is let Y_n be a number of groups of size k in the system and J_n is the remaining number in the system that do not form a group of size k. In order words, $J_n = X_n \ mod \ k$. The transition matrix associated with this DTMC can be written in block form as

$$P = \begin{bmatrix} B & C & & & \\ A_2 & A_1 & A_0 & & \\ & A_2 & A_1 & A_0 & \\ & & \ddots & \ddots & \ddots \end{bmatrix}, \tag{5.19}$$

with

$$B = \begin{bmatrix} p_{0,0} & p_{0,1} & & \\ p_{1,0} & p_{1,1} & p_{1,2} & \\ \vdots & \vdots & \vdots & \ddots \\ p_{k-1,0} & p_{k-1,1} & \cdots & p_{k-1,k-1} \end{bmatrix},$$

$$C = \begin{bmatrix} 0 & 0 & & \\ 0 & 0 & 0 & \\ \vdots & \vdots & \vdots & \ddots \\ p_{k-1,k} & 0 & \cdots & 0 \end{bmatrix},$$

$$A_2 = \begin{bmatrix} c_{k+1} & c_k & \cdots & c_2 \\ & c_{k+1} & \cdots & \vdots \\ & & \ddots & \vdots \\ & & & c_{k+1} \end{bmatrix}, \quad A_0 = \begin{bmatrix} 0 & 0 & & \\ 0 & 0 & 0 & \\ \vdots & \vdots & \vdots & \ddots \\ c_0 & 0 & \cdots & 0 \end{bmatrix},$$

$$A_1 = \begin{bmatrix} c_1 & c_0 & & \\ c_2 & c_1 & c_0 & \\ \vdots & \vdots & \cdots & \ddots \\ c_k & c_{k-1} & \cdots & c_1 \end{bmatrix},$$

where the entries c_j are as given above.

This system is now set up to use the QBD results. Provided the system is stable we have a matrix R which is the minimal non-negative solution to the matrix quadratic equation

$$R = A_0 + RA_1 + R^2A_2.$$

Because of the structure of A_0, i.e. only its bottom left corner element is non-zero, then we have an R matrix which has only the bottom row elements that are non-zero. The R matrix has the following structure

$$R = \begin{bmatrix} 0 & 0 & & \\ 0 & 0 & 0 & \\ \vdots & \vdots & \vdots & \ddots \\ r_0 & r_1 & \cdots & r_{k-1} \end{bmatrix}. \tag{5.20}$$

Now we have the stationary distribution of this system given as

$$\mathbf{y} = \mathbf{y}P, \ \mathbf{y1} = 1,$$

where $\mathbf{y} = [\mathbf{y}_0, \mathbf{y}_1, \ldots,]$, $\mathbf{y}_i = [\mathbf{y}_{i,0}, \mathbf{y}_{i,1}, \ldots, \mathbf{y}_{i,k-1}] = [x_{k(i-1)}, x_{k(i-1)+1}, \ldots, x_{ki-1}]$. It is straightforward to show that

$$r_j = (r_0)^{j+1}, \tag{5.21}$$

where $r_0 = r$ that was calculated from the GI/M/1 version. Hence computing R simply reduces to actually solving for the r in the GI/M/1 system.

From here we apply the boundary conditions as

$$\mathbf{y}_0 = \mathbf{y}_0 B[R], \ \mathbf{y}_0 \mathbf{1} = 1,$$

where $B[R] = B + RA_2$. The normalization equations are then applied to obtain

$$\mathbf{y}_0 (I - R)^{-1} \mathbf{1} = 1.$$

Now we can apply the matrix-geometric result, i.e.

$$\mathbf{y}_{i+1} = \mathbf{y}_i R, \ i \geq 0.$$

Given $\mathbf{y}$ we can determine $\boldsymbol{x}$.

5.1.1.3 An Example

Consider an example with $a = .25, \ b = .65, \ k = 3$, we have $kb = 1.95 > a = .25$. Therefore the system is stable. We proceed with computing the r which is the solution to the polynomial equation

$$r = (.35 + .65r)^3 (.25 + .75r).$$

The minimal non-negative solution to this equation is given as

$$r = 0.0118.$$

Next we compute R matrix. First we compute A_0, A_1 and A_2, as

$$A_0 = \begin{bmatrix} & & \\ & & \\ 0.0107 & 0 & 0 \end{bmatrix}, \ A_1 = \begin{bmatrix} 0.0919 & 0.0107 & \\ 0.2901 & 0.0919 & 0.0107 \\ 0.4014 & 0.2901 & 0.0919 \end{bmatrix},$$

$$A_2 = \begin{bmatrix} 0.206 & 0.4014 & 0.2901 \\ & 0.206 & 0.4014 \\ & & 0.206 \end{bmatrix}.$$

The R matrix is obtained as

$$R = \begin{bmatrix} 0 & 0 & 0 \\ 0 & 0 & 0 \\ 0.0118 & 1.3924 \times 10^{-004} & 1.6430 \times 10^{-006} \end{bmatrix}.$$

We notice that $r_j = r^{j+1}$.

5.1.1.4 Waiting Times

We can study this using direct probability arguments (we call this the traditional approach) or using the absorbing Markov chain idea. Now we work on the basis that the vector $\boldsymbol{x} = \boldsymbol{x}P$, $\boldsymbol{x}\mathbf{1} = 1$ has been solved. We consider the FCFS system. We will only present the waiting time in the queue; the waiting time in the system can be easily inferred from the waiting time in the queue.

We define W^q as the waiting time in the queue for a packet, and let $W_i^q = Pr\{W^q \leq i\}$, $i \geq 0$.

- **Traditional approach:** This approach uses probability arguments directly. It is straightforward to show that

$$W_0^q = \sum_{j=0}^{k-1} x_j. \tag{5.22}$$

The remaining W_j^q, $j \geq 1$ are obtained as follows: Let H_w^{v-k+1} be the probability that it takes no more than w units of time to complete the services of $v-k+1$ packets in the system when there are v packets in the system. Then

$$W_j^q = \sum_{v=k}^{j} H_j^{v-k+1} x_v, \quad j \geq 1. \tag{5.23}$$

The term H_i^j can be obtained recursively as follows:
We know that

$$H_i^1 = 1 - (1-b)^i, \quad i \geq 1, \tag{5.24}$$

then we have

$$H_i^j = \sum_{v=1}^{i-j+1} b^v H_{i-v}^{j-1}, \quad j \leq i \leq \infty. \tag{5.25}$$

Next we develop the z-transform of the waiting time, but before we go further first let us develop some useful transform that will be helpful. Let h_i be defined as the probability that the service time of a packet is no more than i and $h^*(z)$ as the corresponding z-transform, then we have

$$h^*(z) = \sum_{i=1}^{\infty} z^i h_i = \sum_{i=1}^{\infty} z^i (1-(1-b)^i), \quad |z| < 1. \tag{5.26}$$

It is straightforward to see that

$$h^*(z) = \frac{zb}{(1-z)(1-(1-b)z)}, \quad |z| < 1. \tag{5.27}$$

We know that $f^*(z) = \frac{zb}{1-z(1-b)}$ is the z-transform of the distribution of service time of a packet. If we define $h_i^{(j)}$ as the probability that the total service times of j packets is less than or equal to $i \geq j$ then we can easily show that for the case of 2 packets

$$h_i^{(2)} = f_1 h_{i-1} + f_2 h_{i-2} + \cdots + f_{i-1} h_1, \; i \geq 2. \tag{5.28}$$

Taking the z-transform of this and letting it be defined as $h^{(2)}(z)$ we have

$$h^{(2)^*}(z) = h(z)f(z) = \frac{(bz)^2}{(1-z)(1-z(1-b))^2}. \tag{5.29}$$

After repeating this process incrementally we have

$$h^{(j)}(z) = h^{(j-1)}(z)f(z) = \frac{(bz)^j}{(1-z)(1-z(1-b))^j}, \quad |z| < 1. \tag{5.30}$$

Now if we let $W^{q*}(z) = \sum_{v=0}^{\infty} z^v W_v^q, \; |z| < 1$, then we have

$$W^{q*}(z) = \sum_{j=0}^{k-1} x_j + \sum_{v=k}^{\infty} x_v \frac{(bz)^{v-k+1}}{(1-z)(1-z(1-b))^{v-k+1}}, \quad |z| < 1. \tag{5.31}$$

This reduces to

$$W^{q*}(z) = (x_k r^{-1}) \sum_{v=k}^{\infty} \frac{(bzr)^{v-k+1}}{(1-z)(1-z(1-b))^{v-k+1}} + \sum_{j=0}^{k-1} x_j, \quad |z| < 1. \tag{5.32}$$

- **Absorbing Markov chain approach:** This is based on the matrix-analytic approach. Since this is a FCFS system, once an arbitrary packet joins the queue, only the packets ahead of it at that time complete service ahead of it. Hence by studying the DTMC that keeps track of packets ahead of this arbitrary packet we end up with a transition matrix P_w, where

$$P_w = \begin{bmatrix} I & & & \\ \tilde{A}_2 & \tilde{A}_1 & & \\ & \tilde{A}_2 & \tilde{A}_1 & \\ & & \ddots & \ddots \end{bmatrix}, \tag{5.33}$$

with

$$\tilde{A}_2 = \begin{bmatrix} b_k^k & b_{k-1}^k & \cdots & b_1^k \\ & b_k^k & \cdots & b_2^k \\ & & \ddots & \vdots \\ & & & b_k^k \end{bmatrix}, \quad \tilde{A}_1 = \begin{bmatrix} b_0^k & & & \\ b_1^k & b_0^k & & \\ \vdots & \vdots & \ddots & \\ b_{k-1}^k & b_{k-2}^k & \cdots & b_0^k \end{bmatrix}.$$

Let us define a vector $\mathbf{u}^{(0)} = \mathbf{y}$, with corresponding smaller vectors $\mathbf{u}_i^{(0)} = \mathbf{y}_i$, and let

$$\mathbf{u}^{(i)} = \mathbf{u}^{(i-1)} P_w = \mathbf{u}^{(0)} P_w^i, \ i \geq 1.$$

We have

$$W_i^q = Pr\{W^q \leq i\} = \mathbf{u}_0^{(i)} \mathbf{1}, \ i \geq 1, \tag{5.34}$$

with

$$W_0^q = \mathbf{u}_0^{(0)} \mathbf{1}. \tag{5.35}$$

5.1.2 GI/Geo/k Systems

This system can be studied using different techniques. However we are going to present the matrix-analytic approach which is easier to explain and the display of its matrix makes it easy to understand. The matrix approach uses the remaining (residual) interarrival time as a supplementary variable.

We assume that this system has interarrival times that have a general distribution vector $\mathbf{a} = [a_1, \ a_2, \ \cdots]$. The service times follow the geometric distribution with parameter b, and b_i^j is the probability that the services of i out of the j in service are completed in a time epoch. We have $k < \infty$ identical servers in parallel.

5.1.2.1 Using MAM - Same as using supplementary variables

Consider the DTMC $\{(X_n, J_n), n \geq 0\}$, $X_n \geq 0$, $J_n = 1, 2, \cdots$. At time n, we define X_n as the number of packet in the system and J_n as the remaining time before the next packet arrives. The transition matrix representing this DTMC is given as

$$P = \begin{bmatrix} B_{0,0} & B_{0,1} & & & & & \\ B_{1,0} & B_{1,1} & B_{1,2} & & & & \\ \vdots & \vdots & \vdots & \ddots & & & \\ B_{k,0} & B_{k,1} & \cdots & B_{k,k} & B_{k,k+1} & & \\ & A_{k+1} & A_k & A_{k-1} & \cdots & A_0 & \\ & & A_{k+1} & A_k & A_{k-1} & \cdots & A_0 \\ & & & \ddots & \ddots & \ddots & \ddots \ \ddots \end{bmatrix}, \tag{5.36}$$

where the block matrices are defined as follows. First we define new matrices to help us make a compact representation of the blocks. Define I_∞ as an identity matrix of infinite order, $\mathbf{0}_\infty^T$ as an infinite order row vector of zeros and 0_∞ as a infinite order square matrix of zeros. Now we can present the block matrices as

$$A_0 = \begin{bmatrix} b_0^k \mathbf{a} \\ 0_\infty \end{bmatrix}, \ A_{k+1} = \begin{bmatrix} \mathbf{0}_\infty^T \\ b_k^k I_\infty \end{bmatrix}, \ A_i = \begin{bmatrix} b_i^k \mathbf{a} \\ b_{i-1}^k I_\infty \end{bmatrix}, \ 1 \leq i \leq k,$$

$$B_{0,0} = \begin{bmatrix} \mathbf{0}_\infty^T \\ I_\infty \end{bmatrix}, \; B_{0,1} = \begin{bmatrix} \mathbf{a} \\ 0_\infty \end{bmatrix},$$

$$B_{i,0} = \begin{bmatrix} \mathbf{0}_\infty^T \\ b_i^i I_\infty \end{bmatrix}, \; 1 \leq i \leq k, \; B_{i,i+1} = \begin{bmatrix} b_0^i \mathbf{a} \\ 0_\infty \end{bmatrix}, \; 1 \leq i \leq k,$$

$$B_{i,j} = \begin{bmatrix} b_j^i \mathbf{a} \\ b_{j-1}^i I_\infty \end{bmatrix}, \; 1 \leq i \leq k, \; 1 \leq j \leq i \leq k.$$

As an example, we display one of the blocks in detail. Consider the block matrix $A_i, \; 1 \leq i \leq k$ we write out as

$$A_i = \begin{bmatrix} b_i^k a_1 & b_i^k a_2 & b_i^k a_3 & b_i^k a_4 & b_i^k a_5 & \cdots \\ b_{i-1}^k & & & & & \\ & b_{i-1}^k & & & & \\ & & b_{i-1}^k & & & \\ & & & b_{i-1}^k & & \\ & & & & b_{i-1}^k & \\ & & & & & \ddots \end{bmatrix}.$$

We assume that the system is stable and let the $x_{i,j} = Pr\{X_n = i, J_n = j\}|_{n \to \infty}$, and further let $\boldsymbol{x}_i = [x_{i,1}, \; x_{i,2}, \; \cdots]$ and $\boldsymbol{x} = [\boldsymbol{x}_0, \; \boldsymbol{x}_1, \; \cdots]$. Then we know that $\boldsymbol{x}$ is given as the unique solution to

$$\boldsymbol{x} = \boldsymbol{x}P, \;\; \boldsymbol{x}\mathbf{1} = 1.$$

We also know that there is a matrix R which is the non-negative solution to the matrix polynomial

$$R = \sum_{j=0}^{k+1} R^j A_j.$$

This R matrix has the following structure, by virtue of the structure of matrix A_0,

$$R = \begin{bmatrix} r_1 & r_2 & r_3 & \cdots \\ 0 & 0 & 0 & \cdots \\ 0 & 0 & 0 & \cdots \\ \vdots & \vdots & \vdots & \vdots \end{bmatrix}, \qquad (5.37)$$

where $r_i, \; i = 1, 2, \cdots$, are scalar elements.

Clearly, to obtain the matrix R we will have to truncate it at some point. It is intuitive that if the elements of $\mathbf{a} = [a_1, \; a_2, \; a_3, \; \cdots]$ are decreasing from a point $K < \infty$ onwards then the elements of $\mathbf{r} = [r_1, \; r_2, \; r_3, \; \cdots]$ will also be decreasing from around that point. In such situations, which are the ones encountered in practice, we can assume that an appropriate truncation would be acceptable. After obtaining the R matrix we simply go ahead and apply the standard matrix-geometric results to this problem.

Now we consider a special case where the interarrival times have finite support.

Case of finite support for interarrival times:

When the interarrival time has a finite support $K_a = n_t < \infty$, then this problem becomes easier to handle. There is no truncation needed and capitalizating on the structure of A_0 still helps tremendously. First let us define an identity matrix I_j of order j a row vector of zeros $\mathbf{0}_j^T$ of order j and a corresponding column vector $\mathbf{0}_j$. The structure of the transition matrix remains the same, however the blocks need to be defined more specifically for dimension. We have

$$A_0 = \begin{bmatrix} b_0^k \mathbf{a} \\ \mathbf{0}_{n_t-1}\mathbf{0}_{n_t}^T \end{bmatrix}, \; A_{k+1} = \begin{bmatrix} \mathbf{0}_{n_t-1}^T & 0 \\ b_k^k I_{n_t-1} & \mathbf{0}_{n_t-1} \end{bmatrix},$$

$$A_i = \begin{bmatrix} b_i^k \mathbf{a} \\ \tilde{A}_i \end{bmatrix}, \; \tilde{A}_i = [b_{i-1}^k I_{n_t-1}, \; \mathbf{0}_{n_t-1}], \; 1 \le i \le k,$$

$$B_{0,0} = \begin{bmatrix} \mathbf{0}_{n_t-1}^T & 0 \\ I_{n_t-1} & \mathbf{0}_{n_t-1} \end{bmatrix},$$

$$B_{0,1} = \begin{bmatrix} \mathbf{a} \\ \mathbf{0}_{n_t-1}\mathbf{0}_{n_t}^T \end{bmatrix}, \; B_{i,0} = \begin{bmatrix} \mathbf{0}_{n_t-1}^T & 0 \\ b_i^i I_{n_t-1} & \mathbf{0}_{n_t-1} \end{bmatrix}, \; 1 \le i \le k,$$

$$B_{i,i+1} = \begin{bmatrix} b_0^i \mathbf{a} \\ \mathbf{0}_{n_t-1}\mathbf{0}_{n_t}^T \end{bmatrix}, \; 1 \le i \le k,$$

$$B_{i,j} = \begin{bmatrix} b_j^i \mathbf{a} \\ \tilde{B}_{i,j} \end{bmatrix}, \; \tilde{B}_{i,j} = b_{j-1}^i I_{n_t-1}, \; \mathbf{0}_{n_t-1}], \; 1 \le i \le k, \; 1 \le j \le i \le k.$$

5.1.2.2 Numerical Example

Consider the case where $k = 3$, $b = 0.25$. We assume that the interarrival time has a finite support $n_t = 4$, with $\mathbf{a} = [0.1, \; 0.3, \; 0.5, \; 0.1]$. The mean arrival rate is $\lambda = (\sum_{j=1}^{4} j a_j)^{-1} = 0.3846$ and service rate is $\mu = kb = 0.75$. We only display A_0, A_1, A_2, A_3, and A_4.

$$A_0 = \begin{bmatrix} a_1 b_0^3 & a_2 b_0^3 & a_3 b_0^3 & a_4 b_0^3 \\ 0 & 0 & 0 & 0 \\ 0 & 0 & 0 & 0 \\ 0 & 0 & 0 & 0 \end{bmatrix} = \begin{bmatrix} .0016 & .0047 & .0078 & .0016 \\ 0 & 0 & 0 & 0 \\ 0 & 0 & 0 & 0 \\ 0 & 0 & 0 & 0 \end{bmatrix},$$

$$A_1 = \begin{bmatrix} a_1 b_1^3 & a_2 b_1^3 & a_3 b_1^3 & a_4 b_1^3 \\ b_0^3 & 0 & 0 & 0 \\ 0 & b_0^3 & 0 & 0 \\ 0 & 0 & b_0^3 & 0 \end{bmatrix} = \begin{bmatrix} .0141 & .0422 & .0703 & .0141 \\ .0156 & 0 & 0 & 0 \\ 0 & .0156 & 0 & 0 \\ 0 & 0 & .0156 & 0 \end{bmatrix},$$

$$A_2 = \begin{bmatrix} a_1b_2^3 & a_2b_2^3 & a_3b_2^3 & a_4b_2^3 \\ b_1^3 & 0 & 0 & 0 \\ 0 & b_1^3 & 0 & 0 \\ 0 & 0 & b_1^3 & 0 \end{bmatrix} = \begin{bmatrix} .0422 & .1266 & .2109 & .0422 \\ .1406 & 0 & 0 & 0 \\ 0 & .1406 & 0 & 0 \\ 0 & 0 & .1406 & 0 \end{bmatrix},$$

$$A_3 = \begin{bmatrix} a_1b_3^3 & a_2b_3^3 & a_3b_3^3 & a_4b_3^3 \\ b_2^3 & 0 & 0 & 0 \\ 0 & b_2^3 & 0 & 0 \\ 0 & 0 & b_2^3 & 0 \end{bmatrix} = \begin{bmatrix} .0422 & .1266 & .2109 & .0422 \\ .4219 & 0 & 0 & 0 \\ 0 & .4219 & 0 & 0 \\ 0 & 0 & .4219 & 0 \end{bmatrix},$$

$$A_4 = \begin{bmatrix} 0 & 0 & 0 & 0 \\ b_3^3 & 0 & 0 & 0 \\ 0 & b_3^3 & 0 & 0 \\ 0 & 0 & b_3^3 & 0 \end{bmatrix} = \begin{bmatrix} 0 & 0 & 0 & 0 \\ .4219 & 0 & 0 & 0 \\ 0 & .4219 & 0 & 0 \\ 0 & 0 & .4219 & 0 \end{bmatrix}.$$

The resulting R matrix is given as

$$R = \begin{bmatrix} .0017 & .0049 & .0080 & .0016 \\ 0 & 0 & 0 & 0 \\ 0 & 0 & 0 & 0 \\ 0 & 0 & 0 & 0 \end{bmatrix}.$$

5.1.3 PH/PH/k Systems

The PH/PH/k system is a queueing system in which the arrival process is phase type with paramters (α, T) of dimension n_t and service is phase type with parameters (β, S) of order n_s, with $k < \infty$ identical servers in parallel. We will see later that this system is really a reasonably general discrete time single node queue with multiple servers. Several other multiserver queues such as the Geo/Geo/k, GI/Geo/k (with finite support for arrival process) and Geo/G/k (with finite support for service times) are all special cases of this PH/PH/k system.

A very important set of matrices that are critical to this system relate to the number of service completions among ongoing services. To that effect let us study the following matrix sequence $\{B_i^j,\ 0 \le i \le j \le k < \infty\}$. This matrix records the probability of i service completions in a time slot when there are j items in service $(i \le j)$. We present a recursion for computing this matrix sequence. First we re-display the matrix sequence $\{B_j^i,\ 0 \le i \le j < \infty\}$ which was defined in the previous chapter.

$$B_0^1 = S, \quad B_1^1 = (\mathbf{s}\beta),$$

$$B_0^j = B_0^{j-1} \otimes S, \quad B_j^j = B_{j-1}^{j-1} \otimes (\mathbf{s}\beta),$$

$$B_i^j = B_{i-1}^{j-1} \otimes (\mathbf{s}\beta) + B_i^{j-1} \otimes S, \ 1 \le i \le j-1.$$

Let us further consider another matrix sequence $\{\tilde{B}_i^j,\ 0 \le i \le j \le k < \infty\}$, with

$$\tilde{B}_1^1 = \mathbf{s}, \tag{5.38}$$

and

$$\tilde{B}_i^j = \tilde{B}_i^{j-1} \otimes S + B_{i-1}^{j-1} \otimes \mathbf{s},\ 1 \leq i \leq j-1. \tag{5.39}$$

Now consider a DTMC $\{(X_n, L_n, J_n(X_n)), n \geq 0\}$ where at time n, X_n is the number of packets in the system, L_n is the phase of arrival and $J_n(X_n)$ is the set of phases of service of the items in service out of the X_n items in the system. The state space is $\{(0,u) \cup (i,u,v(i)),\ \ i \geq 0, u = 1,2,\cdots,n_t, v(i) = \{1,2,\cdots,n_s\}^{min\{i,k\}}\}$. Note that if there are u items in service then $v(u) = \{1,2,\cdots,n_s\}^v$, because we have to keep track of the phases of service of u items. The transition matrix of this Markov chain can be written as

$$P = \begin{bmatrix} C_{0,0} & C_{0,1} & & & & & & \\ C_{1,0} & C_{1,1} & C_{1,2} & & & & & \\ \vdots & \vdots & \vdots & \ddots & & & & \\ C_{k,0} & C_{k,1} & \cdots & C_{k,k} & C_{k,k+1} & & & \\ & A_{k+1} & A_k & A_{k-1} & \cdots & A_0 & & \\ & & A_{k+1} & A_k & A_{k-1} & \cdots & A_0 & \\ & & & \ddots & \ddots & \ddots & \ddots & \ddots \end{bmatrix}, \tag{5.40}$$

where the block matrices are defined as follows:

$$A_0 = (\mathbf{t}\boldsymbol{\alpha}) \otimes B_0^k,\ A_i = (\mathbf{t}\boldsymbol{\alpha}) \otimes B_i^k + T \otimes B_{i-1}^k,\ 1 \leq i \leq k,$$

$$A_{k+1} = T \otimes B_k^k,\ C_{0,0} = T,\ C_{0,1} = (\mathbf{t}\boldsymbol{\alpha}) \otimes \boldsymbol{\beta},$$

$$C_{i,0} = T \otimes \tilde{B}_i^i,\ C_{i,i+1} = (\mathbf{t}\boldsymbol{\alpha}) \otimes B_0^i \otimes \boldsymbol{\beta},\ 1 \leq i \leq k-1,$$

$$C_{i,v} = T \otimes \tilde{B}_{i-v}^i + (\mathbf{t}\boldsymbol{\alpha}) \otimes \tilde{B}_{i-v+1}^i \otimes \boldsymbol{\beta},\ \ 1 \leq v \leq i \leq k-1,$$

and

$$C_{k,j} = A_{k+1-j},\ \ 0 \leq j \leq k+1.$$

We can now apply matrix-analytic method (MAM) to analyze this system. The procedure is straitghtforward. The only challenge is that of dimensionality. The R matrix is of dimension $n_t {n_s}^k \times n_t {n_s}^k$. This could be huge. As such this problem can only be analyzed using MAM for a moderate size number of servers. For example, if we have an arrival phase type of order $n_t = 2$, service phase of order $n_s = 2$ and number of servers $k = 5$, then our R matrix is of order 64×64, which is easily manageable. However, if the number of servers goes to $k = 10$ we have an R matrix of order 8192×8192 – a big jump indeed.

For a moderate size R we apply the GI/M/1 results of the matrix-geometric approach. We calculate R which is given as the minimal non-negative solution to the equation

$$R = \sum_{v=0}^{k+1} R^v A_v.$$

Given that the system is stable, then we have that

$$\boldsymbol{x} = \boldsymbol{x}P, \ \ \boldsymbol{x}\mathbf{1} = 1,$$

with

$$\boldsymbol{x} = [\boldsymbol{x}_0, \ \boldsymbol{x}_1, \ \boldsymbol{x}_2, \cdots], \ \ \boldsymbol{x}_0 = [x_{0,1}, \ x_{0,2}, \ \cdots, \ x_{0,n_t}],$$

$$\boldsymbol{x}_i = [\boldsymbol{x}_{i,1}, \ \boldsymbol{x}_{i,2}, \ \cdots, \ \boldsymbol{x}_{i,n}], \ \boldsymbol{x}_{i,j} = [x_{i,j,1}, \ x_{i,j,2}, \ \cdots, \ x_{i,j,n_t(i)}],$$

where $n_t(i) = min\{in_s, kn_s\}$.

From the matrix-geometric results we have

$$\boldsymbol{x}_{i+1} = \boldsymbol{x}_i R, \ \ i \geq k.$$

The boundary variables $\boldsymbol{x}_b = [\boldsymbol{x}_0, \ \boldsymbol{x}_1, \ \cdots, \ \boldsymbol{x}_k]$ are obtained by solving

$$\boldsymbol{x}_b = \boldsymbol{x}_b B[R], \ \boldsymbol{x}_b \mathbf{1} = 1,$$

where

$$B[R] = \begin{bmatrix} C_{0,0} & C_{0,1} & & & \\ C_{1,0} & C_{1,1} & C_{1,2} & & \\ \vdots & \vdots & \vdots & \ddots & \\ C_{k-1,0} & C_{k-1,1} & C_{k-1,2} & \cdots & C_{k-1,k} \\ C_{k,0} & C_{k,1} + RA_{k+1} & C_{k,2} + \sum_{v=0}^{1} R^{2-v} A_{k-v+1} & \cdots & C_{k,k+1} + \sum_{v=0}^{k} R^{k-v+1} A_{k-v+1} \end{bmatrix}.$$

This is then normalized by

$$\boldsymbol{x}_0 \mathbf{1} + (\sum_{v=1}^{k-1} \boldsymbol{x}_v)\mathbf{1} + \boldsymbol{x}_k (I-R)^{-1}\mathbf{1} = 1.$$

Usually, all we can manage easily when the dimension of R is huge is just to study the tail behaviour of this system.

5.1.4 Geo/D/k Systems

This type of queue was used extensively in modelling Asynchronous Transfer Mode (ATM) system. The system is as follows. There are $k < \infty$ identical servers in parallel. Packets arrive according to the Bernoulli process with parameter a and each packet requires a constant service time $d < \infty$ which is provided by just one server, any of the k servers, with $k < d$. If we know the total number of packets in the system at time $t \geq 1$, we can easily study the number in the system at times $t, \ t+d, \ t+2d, \ t+3d, \ \cdots, \ t+jd, \ \cdots$. The continuous time analogue of this system is well discussed by Tijms [98]. The first related result on the M/D/k system is by Crommelin [33].

Let $\mathscr{A}_m \geq 0$ be the number of packets that arrive during the time interval of duration $m \geq 1$, and let

$$a_i^m = Pr\{\mathscr{A}_m = i\},\ 0 \leq i \leq m. \tag{5.41}$$

Further let X_n be the number of packets in the system at time n and $c_{w,v} = Pr\{X_{n+1} = v|X_n = w\}|_{n\to\infty}$, then we have

$$c_{w,v} = \begin{cases} a_v^d, & 0 \leq v \leq d;\ 0 \leq w \leq k, \\ a_{v+k-i}^d, & 1 \leq v \leq d;\ i \geq k+1. \end{cases} \tag{5.42}$$

It is clear that X_n is a DTMC with the transition probability matrix P given as

$$P = \begin{bmatrix} a_0^d & a_1^d & a_2^d & \cdots & a_d^d & & & \\ a_0^d & a_1^d & a_2^d & \cdots & a_d^d & & & \\ \vdots & \vdots & \vdots & \cdots & \vdots & & & \\ a_0^d & a_1^d & a_2^d & \cdots & a_d^d & & & \\ & a_0^d & a_1^d & a_2^d & \cdots & a_d^d & & \\ & & a_0^d & a_1^d & a_2^d & \cdots & a_d^d & \\ & & & \ddots & \ddots & \ddots & \ddots & \ddots \end{bmatrix}. \tag{5.43}$$

This P matrix can be re-blocked to have

$$P = \begin{bmatrix} B & C & & & \\ E & A_1 & A_0 & & \\ & A_2 & A_1 & A_0 & \\ & & A_2 & A_1 & A_0 \\ & & & \ddots & \ddots & \ddots \end{bmatrix}, \tag{5.44}$$

where the matrix B is of dimension $k \times k$, C is of dimension $k \times (d-k+1)$, E is of dimension $(d-k+1) \times k$ and $A_j,\ j = 0,1,2$ are each of dimension $(d-k+1) \times (d-k+1)$. We display only B and C and the rest of the block matrices follow.

$$B = \begin{bmatrix} a_0^d & a_1^d & \cdots & a_{k-1}^d \\ a_0^d & a_1^d & \cdots & a_{k-1}^d \\ \vdots & \vdots & \cdots & \vdots \\ a_0^d & a_1^d & \cdots & a_{k-1}^d \end{bmatrix},\quad C = \begin{bmatrix} a_k^d & a_{k+1}^d & \cdots & a_d^d \\ a_k^d & a_{k+1}^d & \cdots & a_d^d \\ \vdots & \vdots & \cdots & \vdots \\ a_k^d & a_{k+1}^d & \cdots & a_d^d \end{bmatrix}.$$

We can now apply the matrix-analytic results for QBD to this problem. We can obtain the matrix R which is the minimal nonnegative solution to the matrix Equation $R = A_0 + RA_1 + R^2A_2$, obtain the solution to $\boldsymbol{x} = \boldsymbol{x}P$, $\boldsymbol{x}\mathbf{1} = 1$ and also solve the boundary equations and normalize. After that we have $\boldsymbol{x}_0$ and $\boldsymbol{x}_1$ determined, where $\boldsymbol{x} = [\boldsymbol{x}_0,\ \boldsymbol{x}_1,\ \boldsymbol{x}_2,\ \cdots]$ and $\boldsymbol{x}_{i+1} = \boldsymbol{x}_i R$.

Finally the distribution of number in the system $x_i = Pr\{X_t = i\}|_{t\to\infty}$ is obtained from the following relationship:

$$\boldsymbol{x}_0 = [x_0,\ x_1,\ \cdots,\ x_{k-1}],\ \boldsymbol{x}_1 = [x_k,\ x_{k+1},\ \cdots,\ x_d],$$

$$\boldsymbol{x}_j = [x_{(j-1)(d-k+1)-k},\ x_{(j-1)(d-k+1)-k+1},\ \cdots,\ x_{j(d-k+1)-k-1}],\ j \geq 2.$$

5.1.4.1 Waiting Times

We can study the waiting time distribution here using a special technique when the service discipline is FCFS. If the service discipline is FCFS, then we can consider this system as one in which we focus only on one server and assign it only the kth arriving packet for service. We thus transform this Geo/D/k system to an NB$_k$/D/1 system for the sake of studying the FCFS waiting time. Here NB$_k$ stands for a negative binomial arrival process with k phases, i.e. $n_t = k$. This NB$_k$ actually captures the arrival process to a target server. Hence this system now has a PH arrival process $(\boldsymbol{\alpha}, T)$ of dimension k with $\boldsymbol{\alpha} = [1,\ 0,\ 0,\ \cdots,\ 0]$ and $T = \begin{bmatrix} 1-a & a & & \\ & 1-a & a & \\ & & \ddots & \ddots \\ & & & 1-a \end{bmatrix}$.

The service time can be represented as a PH distribution $(\boldsymbol{\beta}, S)$ of order $n_s = d$ with $\boldsymbol{\beta} = [1,\ 0,\ 0,\ \cdots,\ 0]$ and $S = \begin{bmatrix} & 1 & & \\ & & 1 & \\ & & & \ddots \\ & & & & 1 \\ & & & & \end{bmatrix}$.

Consider the state space $\Delta = \{(0,\ell) \cup (i,\ell,j),\ i \geq 1, \ell = 1,2,\cdots,k; j = 1,2,\cdots,d\}$, where the first tupple $(0,\ell)$ represents the states with no packet in the system for the target server and arrival is in phase ℓ, i.e. the server is waiting for $(k-\ell)^{th}$ packet to be assigned to it, and the second tupple (i,ℓ,j) represents the states with i packets waiting in the system for the target server, arrival is in phase ℓ and the current packet in service has completed j units of its service. The DTMC representing this system has a transition matrix P of the form

$$P = \begin{bmatrix} B & C & & & \\ E & A_1 & A_0 & & \\ & A_2 & A_1 & A_0 & \\ & & A_2 & A_1 & A_0 \\ & & & \ddots & \ddots & \ddots \end{bmatrix},$$

where

$$B = T,\ C = (\mathbf{t}\boldsymbol{\alpha}) \otimes \boldsymbol{\beta},\ E = T \otimes \mathbf{s},\ A_0 = (\mathbf{t}\boldsymbol{\alpha}) \otimes S,$$

$$A_1 = (\mathbf{t}\boldsymbol{\alpha}) \otimes (\mathbf{s}\boldsymbol{\beta}) + T \otimes S,\ A_2 = T \otimes (\mathbf{s}\boldsymbol{\beta}).$$

By virtue of the structure of matrix A_0 we can see that the associated R matrix for this system has only the last block of rows that is non-zero, i.e. we have

$$R = \begin{bmatrix} 0 & 0 & \cdots & 0 \\ 0 & 0 & \cdots & 0 \\ \vdots & \vdots & \cdots & \vdots \\ 0 & 0 & \cdots & 0 \\ R_1 & R_2 & \cdots & R_{kd} \end{bmatrix}. \tag{5.45}$$

With this structure we can write the equations for the R matrix in block form for each of the kd blocks and save on computations, with

$$R_1 = aS + (1-a)SR_1 + a(\mathbf{s}\boldsymbol{\beta})R_{kd} + (1-a)(\mathbf{s}\boldsymbol{\beta})R_{kd}R_1, \tag{5.46}$$

$$R_j = aSR_{j-1} + (1-a)SR_j + a(\mathbf{s}\boldsymbol{\beta})R_{kd}R_{j-1} + (1-a)(\mathbf{s}\boldsymbol{\beta})R_{kd}R_j,\ 2 \leq j \leq kd. \tag{5.47}$$

From here on we can apply standard QBD results to obtain $\boldsymbol{x} = \boldsymbol{x}P$, $\boldsymbol{x}\mathbf{1} = 1$, where $\boldsymbol{x} = [\boldsymbol{x}_0,\ \boldsymbol{x}_1,\ \cdots]$, and $\boldsymbol{x}_0 = [x_{0,1},\ x_{0,2},\ \cdots,\ x_{0,k}]$, $\boldsymbol{x}_i = [\boldsymbol{x}_{i,1},\ \boldsymbol{x}_{i,2},\ \cdots,\ \boldsymbol{x}_{i,k}]$ and $\boldsymbol{x}_{i,\ell} = [x_{i,\ell,1},\ x_{i,\ell,2},\ \cdots,\ x_{i,\ell,d}]$. Because this is a special case of the PH/PH/1 we simply apply the results of Section 4.9. However, for the Geo/D/k one notices some additional structures which can be explored. For details see Rahman and Alfa [88]. We can now apply the standard results for the PH/PH/1 discussed in Section 4.9 to obtain the waiting times.

5.1.5 MAP/D/k Systems

Ii is straightforward to extend the results of the Geo/D/k to the case of MAP/D/k. We simply change the Bernoulli arrival process with parameters a and $1-a$ to a Markovian arrival process (MAP), with matrices D_1 and D_0, respectively, which is the matrix analogue of the Bernoulli process.

5.1.6 BMAP/D/k Systems

For the more general case with batch Markovian arrival process see Alfa [18].

5.2 Vacation Queues

Vacation queues are a very important class of queues in real life, especially in telecommunication systems where medium access control (MAC) is a critical component of managing a successful network. By definition a vacation queue is a queueing system in which the server is available only a portion of the time. At other times it is busy serving other stations or just not available maybe due to maintenance activities (either routine or due to a breakdown). Polling systems are a class of queueing systems in which a server goes to serve different queues based on a schedule , which implies that the server is not always available to a particular queue. Hence when the server is attending to other queues then it is away on vacation as far as the queue not receiving service is concerned. Vacation queues are used extensively to approximate polling queues.

There are two main categories of vacation queues: gated and ungated systems. In a gated system only the packets in the system waiting at the arrival of a server can get served during that visit. All those that arrive after its return from vacation will not get served during that visit. An ungated system does not have that condition. Under each category there are sub-categories which include exhaustive service, number limited service, time limited service and random vacations. In an exhaustive system all the packets waiting in the system (and only those in the gate for a gated system) will all be served before the server goes on another vacation, i.e. until the queue becomes empty for ungated or the number in the gate is empty for the gated system. For the number limited case the server only processes a pre-determined number of packets during its visit, similarly for time limited the queue is attended to for a fixed length of time. In both cases if the number in the system becomes empty before the limits are reached then the server would go on a vacation. In a random vacation system, the server may stop service at any time and the decision of when to stop serving is random. Generally when a service is interrupted in the time limited and random cases the point of service resumption after a vacation depends on the rule employed; it could be preemptive resume, preemptive repeat, etc. Finally, vacations could be single or multiple. In a single vacation system when the server returns from a vacation it starts to serve immediately or waits for an arrival of packet and then starts to serve. In a multiple vacation system, on the other hand, a server will return to a vacation mode if at the end of a vacation there are no waiting packets. In what follows we discuss mainly ungated systems. We focus on exhaustive systems for the Geo/G/1, GI/Geo/1 and MAP/PH/1 for both single and multiple vacations. We then consider the MAP/PH/1 system for the number limited, time limited and random vacations. We study only the multiple vacations for these last categories.

5.2.1 Geo/G/1 vacation systems

The Geo/G/1 vacation queue is the discrete analogue of the M/G/1 vacation queue that has been used extensively in the literature. In this section we study the exhaustive version of this model for both the single and multiple vacation systems. Consider a single server queue with geometric arrivals with parameter a and general service times S with $b_i = Pr\{S = i\}$, $i = 1, 2, \cdots$, with mean service times b' and b'' is its second moment about zero. The vacation duration V has a general distribution given as $v_i = Pr\{V = i\}$, $i = 1,2,\cdots$, and mean vacation times v' and v'' is its second moment about zero. We will study this system as an imbedded Markov chain.

5.2.1.1 Single vacation system

In a single vacation system, the server goes on a vacation as soon as the system becomes empty. It returns after a random time V and starts to serve the waiting packets or waits to serve the first arriving packet if there is no waiting packets at its arrival. This system can be studied using the imbedded Markov chain idea or using the method of supplementary variables. The imbedded Markov chain approach leads to an M/G/1 type Markov chain and modified methods for the M/G/1 models can be applied. The supplementary variable approach is cumbersome to work with in scalar form but when written in matrix form it leads to a QBD all be it with infinite blocks which will have to be truncated. We focus on the imbedded Markov chain approach.

Imbedded Markov chain approach: At time n, let X_n be the number of packets in the system, with J_n the state of the vacation, with $J_n = 0$ representing vacation termination and $J_n = 1$ representing service completion. Let A_s and A_v be the number of arrivals during a service time and during a vacation period, respectively. We can write

$$(X_{n+1}, J_{n+1}) = \begin{cases} X_n + A_s - 1, 1; X_n \geq 1 \\ A_s, 1; \qquad (X_n, J_n) = (0,0), \\ A_v, 0; \qquad (X_n, J_n) = (0,1) \end{cases} \tag{5.48}$$

Let $a_k^y = Pr\{A_y = k\}, y = s, v$, then

$$a_k^s = \sum_{m=k}^{\infty} b_m \binom{m}{k} a^k (1-a)^{m-k}, \quad k = 0,1,2,\cdots, \text{ and} \tag{5.49}$$

$$a_k^v = \sum_{m=k}^{\infty} v_m \binom{m}{k} a^k (1-a)^{m-k}, \quad k = 0,1,2,\cdots. \tag{5.50}$$

Define $G_k^*(az+1-a) = (az+1-a)^k$, $k \geq 0$ and $A_y^*(z) = \sum_{k=0}^{\infty} a_k^y z^k$, $y = s, v$, $|z| \leq 1$. Further define $B^*(z) = \sum_{k=1}^{\infty} b_k z^k$, $V^*(z) = \sum_{k=1}^{\infty} v_k z^k$, $|z| \leq 1$. We can write

$$A_s^*(z) = \sum_{k=0}^{\infty} z^k \sum_{m=k}^{\infty} b_m \binom{m}{k} a^k (1-a)^{m-k}. \tag{5.51}$$

After routine algebraic operations we have

$$A_s^*(z) = \sum_{k=0}^{\infty} b_k G_k^*(az+1-a), \tag{5.52}$$

$$A_v^*(z) = \sum_{k=0}^{\infty} v_k G_k^*(az+1-a). \tag{5.53}$$

These can be written as

$$A_s^*(z) = B^*(G(az+1-a)), \tag{5.54}$$

$$A_v^*(z) = V^*(G^*(az+1-a)), \ |z| \le 1. \tag{5.55}$$

Let us assume the system is stable and let

$$x_{i,j} = lim_{n\to\infty} Pr\{X_n = i, \ J_n = j\}, \ i = 0,1,2,\cdots,: j = 0,1,$$

then we have the following steady state equations

$$x_{i,0} = x_{0,1} a_i^v, \ \ i = 0,1,2,\cdots \tag{5.56}$$

$$x_{i,1} = x_{0,0} a_i^s + \sum_{k=1}^{i+1} (x_{k,0} + x_{k,1}) a_{i-k+1}^s, \ i = 0,1,2,\cdots. \tag{5.57}$$

Letting

$$x_1^*(z) = \sum_{i=0}^{\infty} x_{i,1} z^j, \text{ and } x^*(z) = \sum_{i=0}^{\infty} (x_{i,0} + x_{i,1}) z^i, \ |z| \le 1,$$

we have

$$x_1^* z = \frac{a_0^v A_s^*(z)(z-1) + A_s^*(z)[A_v^*(z) - 1]}{z - A_s^*(z)} x_{0,1} \tag{5.58}$$

and

$$x^*(z) = \frac{zA_v^*(z) - A_s^*(z) + a_0^v A_s^*(z)(z-1)}{z - A_s^*(z)} x_{0,1}. \tag{5.59}$$

By setting $z = 1$ and using l'Hôpital's rule we obtain

$$1 = x(1) = \frac{1 + a\bar{v} - ab' + a_0^v}{1 - ab'} x0,1, \tag{5.60}$$

$$x_1(1) = \frac{a_0^v + a\bar{v}}{1 - ab'} x_{0,1} = \frac{a_0^v + av'}{1 + av' - ab' + a_0^v}. \tag{5.61}$$

Noting that the mean number in the system $E[X]$ is $(x_1(1))^{-1} \frac{dx_1(z)}{dz}|_{z\to 1}$ we obtain

$$E[X] = ab' + \frac{a^2 b''}{2(1-ab')} + \frac{a^2 v''}{2(a_0^v + av')}. \tag{5.62}$$

This is the well known decomposition result by several authors [42] [36] [37]. By decomposition we mean the result consists of three component with the first two being due to the Geo/G/1 system without a vacation and the third being due to a vacation.

Supplementary variable approach: For the supplementary variable approach we keep track of the number of packets in the system, the remaining time to service completion when the server is busy and the remaining time to the end of a vacation if the server is on a vacation. We will present the supplementary variable approach in detail later for the case of multiple vacations.

5.2.1.2 Multiple vacations system

In this system the server goes on a vacation as soon as the system becomes empty. It returns after a random time V and starts to serve the waiting packets or goes back for another vacation of duration V if there is no waiting packets at its arrival.

Imbedded Markov chain approach: Considering the imbedded approach and using the same notation as in the case of the single vacation we have

$$(X_{n+1}, J_{n+1}) = \begin{cases} X_n + A_s - 1, 1; X_n \geq 1 \\ A_v, 0; \qquad\qquad (X_n, J_n) = (0,1) \end{cases} \tag{5.63}$$

Here the term A_v is the number of arrivals during all the consecutive multiple vacations before the server finds at least one packet in the system. Hence

$$a_j^v = a_j^v (1 - a_0^v)^{-1}, \; j = 1, 2, \cdots.$$

Following the same procedure as in the case of the single vacation we have

$$E[X] = ab' + \frac{a^2 b''}{2(1 - ab')} + \frac{a^2 v''}{2v'}. \tag{5.64}$$

The decomposition result is also apparent here.

Supplementary variable approach: For the supplementary variable approach we keep track of the number of packets in the system, the remaining time to service completion when the server is busy and the remaining time to the end of a vacation if the server is on a vacation. Consider the state space

$$\Delta = \{(0,l) \cup (i,j) \cup (i,l), i \geq 1, l = 1, 2, \cdots; j = 1, 2, \cdots.\}$$

The first pair $(0,l)$ refers to the states when the system is empty and the server is on a vacation and the remaining time to the end of the vacation is l. The second pair (i,j) refers to the states when the server is busy serving with i packets in the system and the remaining service time of the packet in service is j. Finally the third pair (i,l) refers to the states when there are i packets in the system and the server is still on vacation which has l more units of time before completion. The transition matrix P of this DTMC can be represented as

$$P = \begin{bmatrix} B & C & & \\ E & A_1 & A_0 & \\ & A_2 & A_1 & A_0 \\ & & \ddots & \ddots & \ddots \end{bmatrix},$$

where

$$A_k = \begin{bmatrix} A_k^{11} & A_k^{12} \\ A_k^{21} & A_k^{22} \end{bmatrix}, \; k = 0,1,2; C = \begin{bmatrix} C^{11} & C^{12} \end{bmatrix}, \; E = \begin{bmatrix} E^{11} \\ E^{21} \end{bmatrix}, \text{ with}$$

$$A_0^{11} = A_0^{22} = \begin{bmatrix} 0 & 0 & 0 & \cdots \\ a & 0 & \cdots & \\ 0 & a & 0 & \cdots \\ & & \ddots & \end{bmatrix}, \; A_0^{21} = \begin{bmatrix} ab_1 & ab_2 & ab_3 & ab_4 & \cdots \\ 0 & 0 & 0 & 0 & \cdots \\ 0 & 0 & 0 & 0 & \cdots \\ \vdots & \vdots & \vdots & \vdots & \vdots \end{bmatrix}, \; A_0^{12} = 0_\infty,$$

$$A_1^{11} = \begin{bmatrix} ab_1 & ab_2 & ab_3 & \cdots \\ 1-a & 0 & \cdots & \\ 0 & 1-a & 0 & \cdots \\ & & \ddots & \end{bmatrix},$$

$$A_1^{21} = \begin{bmatrix} (1-a)b_1 & (1-a)b_2 & (1-a)b_3 & (1-a)b_4 & \cdots \\ 0 & 0 & 0 & 0 & \cdots \\ 0 & 0 & 0 & 0 & \cdots \\ \vdots & \vdots & \vdots & \vdots & \vdots \end{bmatrix},$$

$$A_1^{22} = \begin{bmatrix} 0 & 0 & 0 & \cdots \\ 1-a & 0 & \cdots & \\ 0 & 1-a & 0 & \cdots \\ & & \ddots & \end{bmatrix}, \; A_1^{12} = 0_\infty,$$

$$A_2^{11} = \begin{bmatrix} (1-a)b_1 & (1-a)b_2 & (1-a)b_3 & (1-a)b_4 & \cdots \\ 0 & 0 & 0 & 0 & \cdots \\ 0 & 0 & 0 & 0 & \cdots \\ \vdots & \vdots & \vdots & \vdots & \vdots \end{bmatrix}, \; A_2^{12} = A_2^{21} = A_2^{22} = 0_\infty,$$

$$B = \begin{bmatrix} (1-a)v_1 & (1-a)v_2 & (1-a)v_3 & \cdots \\ 1-a & 0 & \cdots & \\ 0 & 1-a & 0 & \cdots \\ & & \ddots & \end{bmatrix},$$

$$C^{11} = \begin{bmatrix} av_1 & av_2 & av_3 & \cdots \\ 0 & 0 & \cdots & \\ 0 & 0 & 0 & \cdots \\ & & \ddots & \end{bmatrix}, \; C^{12} = \begin{bmatrix} 0 & 0 & 0 & \cdots \\ a & 0 & \cdots & \\ 0 & a & 0 & \cdots \\ & & \ddots & \end{bmatrix},$$

$$E^{11} = \begin{bmatrix} (1-a)b_1 & (1-a)b_2 & (1-a)b_3 & (1-a)b_4 & \cdots \\ 0 & 0 & 0 & 0 & \cdots \\ 0 & 0 & 0 & 0 & \cdots \\ \vdots & \vdots & \vdots & \vdots & \vdots \end{bmatrix}, \; E^{21} = 0_\infty.$$

We explain three of the block matrices in detail and the rest become straightforward to understand. The matrix B refers to a system that was empty and the server was on vacation and remained on vacation after one time transition with the system still empty. The block matrix C refers to a transition from an empty system to a system with an arrival and it has two components; C^{11} refers to the case when a server was on vacation and vacation ends after one transition and service begins; and C^{12} is the case of the server on vacation and vacation continued after one transition. The block matrix A_0 refers to a transition from a non-empty system with one arrival, leaving the number in the system having increased by one, and it has four components; A_0^{11} refers to a transition from a system in which the server was busy serving and continues to serve after the transition; A_0^{12} refers to a transition from a system in which the server was busy serving and then goes on a vacation after the transition (not possible since the system is not empty); A_0^{21} refers to a transition from a system in which the server was on a vacation and returns to serve after the transition; and A_0^{22} refers to a transition from a system in which the server was on vacation and continues to remain on vacation after the transition.

It is immediately clear that we can apply the matrix-geometric results here. The only challenge we have to deal with is the fact that the block matrices are infinite. If we assume that the system is stable then we can compute the R and G matrices after we truncate the service times and the vacation times distributions. For most practical systems we expect these two quantities to be bounded. So we can go ahead and assume they are bounded. Once we make this assumption applying the results of the QBD is straightforward. First we need to compute the matrix R given by the minimal non-negative solution to the matrix quadratic equation

$$R = A_0 + RA_1 + R^2 A_2.$$

However because of the structure of the matrix A_2 we rather compute the matrix G and then use the result to compute R. The matrix G has a block column of zeros – the left half, because the matrix A_2 has that same structure. So we compute matrix G which is the minimal non-negative solution to the matrix quadratic equation

$$G = A_2 + A_1 G + A_0 G^2,$$

where

$$G = \begin{bmatrix} G_1 & 0 \\ G_2 & 0 \end{bmatrix}. \tag{5.65}$$

All we need to solve for iteratively is G_1 from

$$G_1 = A_2^{11} + A_1^{11} G_1 + A_0^{11} (G_1)^2. \tag{5.66}$$

The matrix G_2 is given as

$$G_2 = A_1^{21}G_1 + A_1^{22}G_2 + A_0^{21}(G_1)^2 + A_0^{22}G_2G_1, \tag{5.67}$$

which we can write as

$$G_2^T = F^T + G_2^T(A_1^{22})^T + G_1^TG_2^T(A_0^{22})^T, \tag{5.68}$$

where $F = A_1^{21}G_1 + A_0^{21}(G_1)^2$ and A^T refers to the transpose of the matrix A. The matrix G_2 is then solved for from the linear matrix equation

$$vecG_2^T = vecF^T + (A_1^{22} \otimes I)vecG_2^T + (A_0^{22} \otimes G_1)vecG_2^T, \tag{5.69}$$

simply as

$$vecG_2^T = vecF^T(I - (A_1^{22} \otimes I) - (A_0^{22} \otimes G_1))^{-1}. \tag{5.70}$$

5.2.2 GI/Geo/1 vacation systems

The GI/Geo/1 vacation queue is easily studied using either the imbedded Markov chain approach or the supplementary variable technique. Tian and Zhang [97] used a combination of the imbedded Markov chain and matrix-geometric approach to analyze this system, similar in principle to the case of the Geo/G/1 vacation queue, except for a minor difference. They studied the system at points of arrivals, as expected, however in addition they distinguished between whether an arrival occurs when the server was busy or when it was on a vacation. Whereas in the case of the GI/Geo/1 vacation the system was studied at departure epochs and at return times from vacations. Details of this GI/Geo/1 approach can be found in that paper by Tian and Zhong [97]. In this book we consider the supplementary variables technique only and show how the matrix-geometric methods can be used to analyze this system. This is presented only for the multiple vacation system. The case of single vacation can be easily inferred from the results provided.

Supplementary variable technique and MGM: We consider the multiple vacation system. Packets arrive with general interarrival times $\mathscr{A}$ and $a_i = Pr\{\mathscr{A} = i\}, i = 1, 2, \cdots$. The service times follow the geometric distribution with parameter b and the vacation duration $\mathscr{V}$ has a general distribution with $v_j = Pr\{\mathscr{V} = j\}$, $j = 1, 2, \cdots$.

The state space for this system can be written as

$$\Delta = \{(0,l,k) \cup (i,k) \cup (i,l,k)\}, l = 1,2,\cdots; i \geq 1; k = 1,2,\cdots.$$

The first tupple $(0,l,k)$ refers to an empty system, with server on a vacation which has l more units of time to be completed and the next packet arrival is k units of time away. The second tupple (i,k) refers to the case with i packets in the system with the server in service and attending to a packet and the next packet arrival is k units of time away. The final tupple (i,l,k) refers to having i packets in the system, with the server on a vacation which has l more units of time before it is completed and k is the remaining time before the next packet arrives into the system. It is immediately clear that the transition matrix P of this DTMC is a QBD type with

$$P = \begin{bmatrix} B & C & & & \\ E & A_1 & A_0 & & \\ & A_2 & A_1 & A_0 & \\ & & \ddots & \ddots & \ddots \end{bmatrix}.$$

If we assume the interarrival time and the vacation duration to have finite supports or can be truncated (an essential assumption required for algorithmic tractability) then we can represent the interarrival time and also vacation duration by PH distributions. Keeping in mind that a PH is a special case of MAP and geometric distribution is a special case of PH, we can see that any result for a MAP/PH/1 system with PH vacation can be applied to this GI/Geo/1 vacation system. Hence we skip further discussion of the GI/Geo/1 vacation queue and go straight to develop the MAP/PH/1 vacation model in the next section. The only additional feature is that the PH distributions encountered in the GI/Geo/1 vacation have special structures that can be capitalized on. This is left to the reader to explore.

5.2.3 MAP/PH/1 vacation systems

The model for the MAP/PH/1 with PH vacations, (ϕ, V), is sufficient to cover most of the commonly encountered vacation queues. We will develop this model and point out how it can be used to obtain results for other related vacation models. We let the arrival rate be $\lambda = \boldsymbol{\delta} D\mathbf{1}$, where $\boldsymbol{\delta} = \boldsymbol{\delta} D$, $\boldsymbol{\delta}\mathbf{1} = 1$, the mean service time be $\mu^{-1} = \boldsymbol{\beta}(I-S)^{-1}\mathbf{1}$ and the mean vacation duration be $v^{-1} = \phi(I-V)^{-1}\mathbf{1}$.

Consider a system with MAP arrivals represented by two $n_t \times n_t$ matrices D_0 and D_1. As before we have the elements $(D_k)_{i,j}$ representing transitions from state i to state j with $k = 0, 1$, arrivals. The matrix $D = D_0 + D_1$ is stochastic and irreducible. There is a single server that provides a service with PH distribution represented by $(\boldsymbol{\beta}, S)$ of dimension n_s. We assume that the server takes vacations that have PH durations with representation (ϕ, V) of order n_v. We will be specific about the type of vacation as the models are developed.

5.2.3.1 Exhaustive single vacation:

In this system the server goes on a vacation as soon as the system is empty. After it returns from a vacation it starts to serve the waiting packets or waits for the first arriving packet if there are no waiting packets. Consider the DTMC

$\{(X_n, U_n, Z_n, Y_n), n \geq 0\}$, where at time n, X_n is the number of packets waiting in the system, U_n is the arrival phase of the packets, Z_n is the phase of service when the server is in service, and Y_n is the phase of vacation when the server is on a vacation. Essentially, the state space can be written as

$$\Delta = \{(0,k,0) \cup (0,k,l) \cup (i,k,l) \cup (i,k,j);$$

$$k = 1,2,\cdots,n_t; l = 0,1,\cdots,n_v; i = 1,2,\cdots; j = 1,2,\cdots,n_s\}.$$

Here

- The first tuple $(0,k,0)$ refers to an empty system, with arrival in phase k and the server is waiting to serve any arriving packet (i.e. the server is back from a vacation and no waiting packets).
- The second tuple $(0,k,l)$ refers to an empty system, arrival in phase k and server is on vacation which is in phase l.
- The third tuple (i,k,l) refers to having i packets in the system, arrival is in phase k and the server is on vacation which is in phase l.
- The last tuple (i,k,j) refers to having i packets in the system, arrival is in phase k and the server is providing service to packets and the service is in phase j.

The transition matrix P for this DTMC can be written as

$$P = \begin{bmatrix} B & C & & & \\ E & A_1 & A_0 & & \\ & A_2 & A_1 & A_0 & \\ & & A_2 & A_1 & A_0 \\ & & & \ddots & \ddots & \ddots \end{bmatrix},$$

where

$$B = \begin{bmatrix} D_0 & 0 \\ D_0 \otimes \mathbf{v} & D_0 \otimes V \end{bmatrix}, \quad C = \begin{bmatrix} D_1 \otimes \boldsymbol{\beta} & 0 \\ D_1 \otimes (\mathbf{v}\boldsymbol{\beta}) & D_1 \otimes V \end{bmatrix},$$

$$E = \begin{bmatrix} 0 & D_0 \otimes (\mathbf{s}\boldsymbol{\phi}) \\ 0 & 0 \end{bmatrix}, \quad A_2 = \begin{bmatrix} D_0 \otimes (\mathbf{s}\boldsymbol{\beta}) & 0 \\ 0 & 0 \end{bmatrix},$$

$$A_0 = \begin{bmatrix} D_1 \otimes S & 0 \\ D_1 \otimes (\mathbf{v}\boldsymbol{\beta}) & D_1 \otimes V \end{bmatrix}, \quad A_1 = \begin{bmatrix} D_0 \otimes S + D_1 \otimes (\mathbf{s}\boldsymbol{\beta}) & 0 \\ D_0 \otimes (\mathbf{v}\boldsymbol{\beta}) & D_0 \otimes V \end{bmatrix},$$

and $\mathbf{v} = \mathbf{1} - V\mathbf{1}$.

Now we can apply the matrix-geometric method to analyze this system. We know that there is a matrix R which is the minimal non-negative solution to the matrix quadratic equation

$$R = A_0 + RA_1 + R^2 A_2,$$

and the spectral radius of this matrix R is less than one if the system is stable. The system is stable if

$$\boldsymbol{\pi} A_0 \mathbf{1} < \boldsymbol{\pi} A_2 \mathbf{1},$$

where $\boldsymbol{\pi} A = \boldsymbol{\pi}$, $\boldsymbol{\pi}\mathbf{1} = 1$ and $A = A_0 + A_1 + A_2$.

In this particular case we can exploit the structure of the block matrices and compute the matrix G first which is the minimal non-negative solution to the matrix quadratic equation

$$G = A_2 + A_1 G + A_0 G^2.$$

We chose to compute the matrix G first because the matrix A_2 has a block column of zeros which implies that the matrix G is of the form

$$G = \begin{bmatrix} G_1 & 0 \\ G_2 & 0 \end{bmatrix}. \tag{5.71}$$

By writing the G matrix equations in block form we only need to compute $n_t n_s \times n_t n_s + n_t n_r \times n_t n_s$ elements instead of $(n_t n_s + n_t n_r) \times (n_t n_s + n_t n_r)$ elements – a huge saving indeed. The block elements of G can be computed from

$$G_1 = D_0 \otimes (\mathbf{s}\boldsymbol{\beta}) + (D_0 \otimes S + D_1 \otimes (\mathbf{s}\boldsymbol{\beta}))G_1 + (D_1 \otimes S)(G_1)^2, \tag{5.72}$$

$$G_2 = (D_0 \otimes (\mathbf{v}\boldsymbol{\beta}))G_1 + (D_0 \otimes V)G_2 + (D_1 \otimes (\mathbf{v}\boldsymbol{\beta}))(G_1)^2 + (D_1 \otimes V)G_2 G_1. \tag{5.73}$$

We note that G_1 is actually the G matrix of a MAP/PH/1 queue with no vacation. So that is easy to obtain. After solving for G_1 we can simply obtain G_2 in a matrix linear equation by using the vectorization transformation. Let us write the second block of equation as

$$G_2 = F_1 G_1 + F_2 G_2 + F_3 (G_1)^2 + F_4 G_2 G_1. \tag{5.74}$$

Since G_1 is already known we can actually write this equation as

$$G_2 = H_1 + F_2 G_2 + F_4 G_2 F_5, \tag{5.75}$$

where $H_1 = F_1 G_1 + F_3 (G_1)^2$ and $F_5 = G_1$. Consider an $M \times N$ matrix

$$Y = [Y_{.1},\ Y_{.2},\ \cdots,\ Y_{.j},\ \cdots,\ Y_{.N}],$$

where $Y_{.j}$ is the jth column of the matrix Y, we define a $1 \times NM$ column vector

$$vecY = \begin{bmatrix} Y_{.1} \\ Y_{.2} \\ \vdots \\ Y_{.j} \\ \vdots \\ Y_{.N} \end{bmatrix},$$

then we can write the equation for matrix G_2 as

$$vecG_2 = vecH_1 + ((F_5)^T \otimes F_4) vecG_2 \tag{5.76}$$

which leads to

$$vecG_2 = vecH_1(I-((F_5)^T \otimes F_4))^{-1}. \tag{5.77}$$

We assume the inverse of $(I-((F_5)^T \otimes F_4))$ exists. After obtaining the G matrix we can then use it to compute the matrix R using Equation (2.100), i.e.

$$R = A_0(I - A_1 - A_0G)^{-1}.$$

5.2.3.2 Stationary Distribution:

Let the steady state vector of this DTMC be $\boldsymbol{x} = [\boldsymbol{x}_0, \boldsymbol{x}_1, \boldsymbol{x}_2, \cdots,]$, with each of the vectors $\boldsymbol{x}_i$ appropriately defined to match the state space described earlier, then we have

$$\boldsymbol{x} = \boldsymbol{x}P, \ \boldsymbol{x}\mathbf{1} = 1,$$

and

$$\boldsymbol{x}_{i+1} = \boldsymbol{x}_i R, \ i \geq 1.$$

We apply the standard matrix-geometric method to obtain the boundary vectors $[\boldsymbol{x}_0, \boldsymbol{x}_1]$ from

$$[\boldsymbol{x}_0, \boldsymbol{x}_1] = [\boldsymbol{x}_0, \boldsymbol{x}_1]\begin{bmatrix} B & C \\ E & A_1 + RA_2 \end{bmatrix}, \ [\boldsymbol{x}_0, \boldsymbol{x}_1]\mathbf{1} = 1,$$

and then normalize with

$$\boldsymbol{x}_0\mathbf{1} + \boldsymbol{x}_1[I-R]^{-1}\mathbf{1} = 1.$$

From here on the steps required to obtain the expected number in the system and waiting times are straightforward.

5.2.3.3 Example of the Geo/Geo/1 vacation:

Consider the special case of Geo/Geo/1 with geometric vacation. We let arrival have parameter a, service parameter b and vacation parameter c, then our block matrices will be

$$B = \begin{bmatrix} 1-a & 0 \\ (1-a)c & (1-a)(1-c) \end{bmatrix}, \ C = \begin{bmatrix} a & 0 \\ ac & a(1-c) \end{bmatrix},$$

$$E = \begin{bmatrix} 0 & (1-a)b \\ 0 & 0 \end{bmatrix}, \ A_2 = \begin{bmatrix} (1-a)b & 0 \\ 0 & 0 \end{bmatrix},$$

$$A_0 = \begin{bmatrix} a(1-b) & 0 \\ ac & a(1-c) \end{bmatrix}, \ A_1 = \begin{bmatrix} (1-a)(1-b)+ab & 0 \\ (1-a)c & (1-a)(1-c) \end{bmatrix}.$$

For this example, the G matrix is

$$G = \begin{bmatrix} 1 & 0 \\ 1 & 0 \end{bmatrix}.$$

Hence it is simple to show that the R matrix is explicit, and can be written as

$$R = \frac{1}{\bar{a}b(1-\overline{ac})} \begin{bmatrix} a\bar{b}(1-\overline{ac}) & 0 \\ ac(1-\overline{ac}) + a\bar{c}(\bar{a}c+a) & a\overline{ac}\bar{b} \end{bmatrix},$$

where $\bar{a} = 1-a,\ \bar{b} = 1-b,\ \bar{c} = 1-c.$

5.2.3.4 Exhaustive multiple vacations

In this system the server goes on a vacation as soon as the system is empty. After it returns from a vacation it starts to serve the waiting packets but if there are no waiting packets then it proceeds to another vacation. Consider the DTMC $\{(X_n, U_n, Z_n, Y_n), n \geq 0\}$, where at time n, X_n is the number of packets waiting in the system, U_n is the arrival phase of the packets, Z_n is the phase of service when the server is in service, and Y_n is the phase of vacation when the server is on a vacation. Essentially, the state space can be written as

$$\Delta = \{(0,k,l) \cup (i,k,l) \cup (i,k,j);$$

$$k = 1,2,\cdots,n_t; l = 0,1,\cdots,n_v; i = 1,2,\cdots; j = 1,2,\cdots,n_s\}.$$

Here

- The first tuple $(0,k,l)$ refers to an empty system, arrival in phase k and server is on vacation which is in phase l.
- The second tuple (i,k,l) refers to having i packets in the system, arrival is in phase k and the server is on vacation which is in phase l.
- The third tuple (i,k,j) refers to having i packets in the system, arrival is in phase k and the server is providing service to packets and the service is in phase j.

It is immediately clear that the only difference between this multiple vacations and single vacation is at the boundary. We no longer have a server waiting for packets if when it returns from a vacation it finds no waiting packets. The transition matrix P for this DTMC can be written as

$$P = \begin{bmatrix} B & C & & & \\ E & A_1 & A_0 & & \\ & A_2 & A_1 & A_0 & \\ & & A_2 & A_1 & A_0 \\ & & & \ddots & \ddots & \ddots \end{bmatrix},$$

where

$$B = D_0 \otimes (\mathbf{v}\phi + V),\ C = \begin{bmatrix} D_1 \otimes (\mathbf{v}\boldsymbol{\beta}) & D_1 \otimes V \end{bmatrix},$$

$$E = \begin{bmatrix} 0 \\ D_0 \otimes (\mathbf{s}\phi) \end{bmatrix}, \; A_2 = \begin{bmatrix} D_0 \otimes (\mathbf{s}\boldsymbol{\beta}) & 0 \\ 0 & 0 \end{bmatrix},$$

$$A_0 = \begin{bmatrix} D_1 \otimes S & 0 \\ D_1 \otimes (\mathbf{v}\boldsymbol{\beta}) & D_1 \otimes V \end{bmatrix}, \; A_1 = \begin{bmatrix} D_0 \otimes S + D_1 \otimes (\mathbf{s}\boldsymbol{\beta}) & 0 \\ D_0 \otimes (\mathbf{v}\boldsymbol{\beta}) & D_0 \otimes V \end{bmatrix}.$$

Now we can apply the matrix-geometric method to analyze this system. We know that there is a matrix R which is the minimal non-negative solution to the matrix quadratic equation

$$R = A_0 + RA_1 + R^2 A_2,$$

and the spectral radius of this matrix R is less than one if the system is stable. The system is stable if

$$\boldsymbol{\pi} A_0 \mathbf{1} < \boldsymbol{\pi} A_2 \mathbf{1},$$

where $\boldsymbol{\pi} A = \boldsymbol{\pi}$, $\boldsymbol{\pi}\mathbf{1} = 1$ and $A = A_0 + A_1 + A_2$.

Other than the boundary conditions that are different every other aspects of the matrices for the multiple vacations is the same as those for the single vacation. As such we will not repeat the analysis here.

5.2.4 MAP/PH/1 vacation queues with number limited service

Polling systems are very important in telecommunications especially when it comes to medium access control. The server attends to each traffic flow (queue) by serving a predetermined number of packets and then goes to another queue and so on. If such a system is approximated by a vacation model then we have a number limited system. This is a system in which a server attends to a queue and serves up to a predetermined number of packets $\mathcal{K} < \infty$ during each visit and then goes on vacation after this number has been served or the system becomes empty (whichever occurs first). If when the server returns from a vacation the system is empty then it proceeds on another vacation, i.e. it is a multiple vacation rule. This kind of queue is easier to study using the matrix-analytic method than standard methods and as such we present that approach.

Note the variable K is used here different from the previous chapters. This is to reduce the number of notations used in the book. Some variables may be used to denote different parameters when it will not cause any confusion to do so, and hence they will be re-defined appropriately as the need arises.

Consider a MAP/PH/1 vacation queue, where the vacation duration follows a PH distribution with parameters (ϕ, V) of dimension n_v. The arrival follows a MAP and is represented by two matrices D_0 and D_1 of dimension n_t and service is PH with parameters $(\boldsymbol{\beta}, S)$ of dimension n_s. Now consider the following state space

$$\Delta = \{(0,k,l) \cup (i,k,l) \cup (i,u,k,j)\}, k = 1,2,\cdots,n_t; l = 1,2,\cdots,n_v;$$

$$i = 1,2,\cdots; u = 0,1,2,\cdots,K; j = 1,2,\cdots,n_s.$$

Throughout this state space description, the variable k stands for the phase of arrival of the MAP, l the phase of a vacation, j the phase of service, i the number of packets in the system and u the number of packets that have been served during a particular visit to the queue by the server. The first tupple $(0,k,l)$ represents an empty system with the server on a vacation. The second tupple (i,k,l) represents the case when the server is on a vacation and there are i packets waiting in the system and the third tupple (i,u,k,j) represents the case when the server is attending to the queue and there are i packets in the system, with u packets already served. It is immediately clear that the transition matrix P for this DTMC is a QBD type written as

$$P = \begin{bmatrix} B & C & & \\ E & A_1 & A_0 & \\ & A_2 & A_1 & A_0 \\ & & \ddots & \ddots & \ddots \end{bmatrix},$$

where

$$A_0 = \begin{bmatrix} (\mathbf{t}\boldsymbol{\alpha}) \otimes V & \mathbf{e}_1^T(\mathcal{K}) \otimes (\mathbf{t}\boldsymbol{\alpha}) \otimes (\mathbf{v}\boldsymbol{\beta}) \\ 0 & I_{\mathcal{K}} \otimes (\mathbf{t}\boldsymbol{\alpha}) \otimes S \end{bmatrix},$$

$$A_2 = \begin{bmatrix} 0 & 0 \\ \mathbf{e}_{\mathcal{K}}(\mathcal{K}) \otimes T \otimes \boldsymbol{\phi} \otimes \mathbf{s} & \bar{I}(\mathcal{K}-1) \otimes T \otimes (\mathbf{s}\boldsymbol{\beta}) \end{bmatrix},$$

$$A_1 = \begin{bmatrix} T \otimes V & \mathbf{e}_1^T(\mathcal{K}) \otimes T \otimes (\mathbf{v}\boldsymbol{\beta}) \\ \mathbf{e}_{\mathcal{K}}(\mathcal{K}) \otimes (\mathbf{t}\boldsymbol{\alpha}) \otimes \boldsymbol{\phi} \otimes \mathbf{s} & I_{\mathcal{K}} \otimes T \otimes S + \bar{I}(\mathcal{K}-1) \otimes (\mathbf{t}\boldsymbol{\alpha}) \otimes (\mathbf{s}\boldsymbol{\beta}) \end{bmatrix},$$

$$B = T \otimes V,\ C = \begin{bmatrix} (\mathbf{t}\boldsymbol{\alpha}) \otimes V & \mathbf{e}_1^T(\mathcal{K}) \otimes (\mathbf{t}\boldsymbol{\alpha}) \otimes (\mathbf{v}\boldsymbol{\beta}) \end{bmatrix},$$

$$E = \begin{bmatrix} 0 \\ \mathbf{1}(\mathcal{K}) \otimes T \otimes \boldsymbol{\phi} \otimes \mathbf{s} \end{bmatrix},$$

where I_j is an identity matrix of dimension j and $\bar{I}(j) = \begin{bmatrix} 0 & I_j \\ 0 & 0 \end{bmatrix}$, $\mathbf{1}(j)$ is a column vector of ones of dimension j, $\mathbf{e}_j$ is the jth column of an identity matrix and $\mathbf{e}_j^T$ is its transpose; the dimensions are left out to avoid complex notations, but the dimensions are obvious.

We see that this problem is nicely set up as QBD and the results from QBD can be used to analyze this system. We first obtain the associated R matrix and then proceed as discussed in Chapter 2. However, in obtaining the R matrix here we notice that there is no obvious simple structure that can be exploited, hence it is more efficient to use the quadratically converging algorithms such as the Logarithmic Reduction [65] or the Cyclic Reduction [76].

5.2.5 MAP/PH/1 vacation queues with time limited service

In some medium access control protocols the server attends to each traffic flow (queue) for a limited length of time and then goes to another queue and so on. If such a system is approximated by a vacation model then we have a time limited system. This is a system in which a server attends to a queue and serves up to a predetermined length of time $\mathscr{T} < \infty$ during each visit and then goes on vacation after this time has expired or the system becomes empty (whichever occurs first). The server may interrupt the service of a packet that is in service once the time limit is reached. We define a stochastic matrix Q whose elements $Q_{j_1^{'},j_1}$ refer to the probability that a packet's service resumes in phase j_1 after the server returns from a vacation, given that the service was interrupted at phase $j_1^{'}$ at the start of a vacation. If when the server returns from a vacation the system is empty then it proceeds on another vacation, i.e. it is a multiple vacation rule. This kind of queue is also easier to study using the matrix-analytic method.

Consider a MAP/PH/1 vacation queue, where the vacation duration follows a PH distribution with parameters (ϕ, V) of dimension n_v. The arrival follows a MAP and is represented by two matrices D_0 and D_1 of dimension n_t and service is PH with parameters $(\boldsymbol{\beta}, S)$ of dimension n_s. Now consider the following state space

$$\Delta = \{(0,k,l) \cup (i,k,l,j^{'}) \cup (i,u,k,j)\}, k = 1,2,\cdots,n_t; l = 1,2,\cdots,n_v;$$

$$j^{'} = 0,1,2,\cdots,n_s; i = 1,2,\cdots; u = 0,1,2,\cdots,\mathscr{T}; j = 1,2,\cdots,n_s.$$

Throughout this state space description, the variable k stands for the phase of arrival of the MAP, l the phase of a vacation, j the phase of service, $j^{'}$ the phase at which a packet's service was interrupted (this phase could be zero if no interruption took place), i the number of packets in the system and u the time clock recording the time that has expired since during a particular visit to the queue by the server. The first tupple $(0,k,l)$ represents an empty system with the server on a vacation. The second tupple (i,k,l) represents the case when the server is on a vacation and there are i packets waiting in the system and the third tupple (i,u,k,j) represents the case when the server is attending to the queue and there are i packets in the system, with u packets already served. It is immediately clear that the transition matrix P for this DTMC is a QBD type written as

$$P = \begin{bmatrix} B & C & & & \\ E & A_1 & A_0 & & \\ & A_2 & A_1 & A_0 & \\ & & \ddots & \ddots & \ddots \end{bmatrix},$$

where

$$A_0 = \begin{bmatrix} (\mathbf{t}\boldsymbol{\alpha}) \otimes V \otimes I_{n_s+1} & \mathbf{e}_1^T(\mathscr{T}) \otimes (\mathbf{t}\boldsymbol{\alpha}) \otimes \mathbf{v} \otimes Q^* \\ \mathbf{e}_{\mathscr{T}}(\mathscr{T}) \otimes (\mathbf{t}\boldsymbol{\alpha}) \otimes \phi \otimes S^* & I(\mathscr{T}-1) \otimes (\mathbf{t}\boldsymbol{\alpha}) \otimes S \end{bmatrix},$$

$$A_2 = \begin{bmatrix} 0 & 0 \\ \mathbf{e}_{\mathcal{T}}(\mathcal{T}) \otimes T \otimes \phi \otimes (\mathbf{s}\mathbf{e}_1^T(n_s+1)) & \bar{I}(\mathcal{T}-1) \otimes T \otimes (\mathbf{s}\boldsymbol{\beta}) \end{bmatrix},$$

$$A_1 = \begin{bmatrix} T \otimes V \otimes I_{n_s+1} & \mathbf{e}_1^T(\mathcal{T}) \otimes T \otimes \mathbf{v} \otimes Q^* \\ A_1^1 & A_1^2 \end{bmatrix},$$

$$A_1^1 = \mathbf{e}_{\mathcal{T}}(\mathcal{T}) \otimes [(\mathbf{t}\boldsymbol{\alpha}) \otimes \phi \otimes (\mathbf{s}\mathbf{e}_1^T(n_s+1)) + T \otimes \phi \otimes S^*],$$

$$A_1^2 = \bar{I}(\mathcal{T}-1) \otimes (T \otimes S + (\mathbf{t}\boldsymbol{\alpha}) \otimes (\mathbf{s}\boldsymbol{\beta}),$$

$$B = T \otimes V,\ C = \begin{bmatrix} (\mathbf{t}\boldsymbol{\alpha}) \otimes V & \mathbf{e}_1^T(\mathcal{T}) \otimes (\mathbf{t}\boldsymbol{\alpha}) \otimes (\mathbf{v}\boldsymbol{\beta}) \end{bmatrix},$$

$$E = \begin{bmatrix} 0 \\ \mathbf{1}(\mathcal{T}) \otimes T \otimes \phi \otimes \mathbf{s} \end{bmatrix},$$

where I_j is an identity matrix of dimension j and $\bar{I}(j) = \begin{bmatrix} 0 & I_j \\ 0 & 0 \end{bmatrix}$, $\mathbf{1}(j)$ is a column vector of ones of dimension j, $\mathbf{e}_j$ is the jth column of an identity matrix and $\mathbf{e}_j^T$ is its transpose, $S^* = [\mathbf{0}\ \ S]$ an $(n_s+1) \times n_s$ matrix, $Q^* = \begin{bmatrix} \boldsymbol{\beta} \\ Q \end{bmatrix}$, an $n_s \times (n_s+1)$ matrix. All the zeros in the matrices are dimensioned accordingly; the dimensions are left out to avoid complex notations, but the dimensions are obvious.

We see that this problem is nicely set up as QBD and the results from QBD can be used to analyze this system. We first obtain the associated R matrix and then proceed as discussed in Chapter 2. However, in obtaining the R matrix here we notice that there is no obvious simple structure that can be exploited, hence it is more efficient to use the quadratically converging algorithms such as the Logarithmic Reduction [65] or the Cyclic Reduction [76].

5.2.6 *Random Time Limited Vacation Queues/Queues with Server Breakdowns and Repairs*

In this section we study a MAP/PH/1 vacation queue with random time limited visits. In this system the server attends to the queue for a random amount of time and then proceeds on a vacation which is of a random time duration also. The visit to a queue has a PH distribution (γ, U) of order κ. With this random time visits the server may interrupt the service of a packet that is in service once a vacation is set to start. We define a stochastic matrix Q whose elements Q_{j_1', j_1} refers to the probability that a packet's service resumes in phase j_1 after server returns from a vacation, given that the service was interrupted at phase j_1' at the start of a vacation. If when the server returns from a vacation the system is empty then it proceeds on another vacation, i.e. it is a multiple vacation rule. This kind of queue is also easier to study using the matrix-analytic method.

Consider a MAP/PH/1 vacation queue, where the vacation duration follows a PH distribution with parameters (ϕ, V) of dimension n_v. The arrival follows a MAP and is represented by two matrices D_0 and D_1 of dimension n_t and service is PH with parameters $(\boldsymbol{\beta}, S)$ of dimension n_s. Now consider the following state space

$$\Delta = \{(0,k,l) \cup (i,k,l,j') \cup (i,u,k,j)\}, k = 1,2,\cdots,n_t; l = 1,2,\cdots,n_v;$$

$$j' = 0,1,2,\cdots,n_s; i = 1,2,\cdots; u = 0,1,2,\cdots,\kappa; j = 1,2,\cdots,n_s.$$

Throughout this state space description, the variable k stands for the phase of arrival of the MAP, l the phase of a vacation, j the phase of service, j' the phase at which a packet's service was interrupted (this phase could be zero if no interruption took place), i the number of packets in the system and u is the phase of a particular visit to the queue by the server (this is like the time clock that keeps track of how long a system is being attended to). The first tupple $(0,k,l)$ represents an empty system with the server on a vacation. The second tupple (i,k,l) represents the case when the server is on a vacation and there are i packets waiting in the system and the third tupple (i,u,k,j) represents the case when the server is attending to the queue and there are i packets in the system, with u as the phase of a particular visit - time clock. It is immediately clear that the transition matrix P for this DTMC is a QBD type written as

$$P = \begin{bmatrix} B & C & & \\ E & A_1 & A_0 & \\ & A_2 & A_1 & A_0 \\ & & \ddots & \ddots & \ddots \end{bmatrix},$$

where

$$A_0 = \begin{bmatrix} (\mathbf{t}\boldsymbol{\alpha}) \otimes V \otimes I_{n_s+1} & \gamma \otimes (\mathbf{t}\boldsymbol{\alpha}) \otimes \mathbf{v} \otimes Q^* \\ \mathbf{u} \otimes (\mathbf{t}\boldsymbol{\alpha}) \otimes \phi \otimes S^* & U \otimes (\mathbf{t}\boldsymbol{\alpha}) \otimes S \end{bmatrix},$$

$$A_2 = \begin{bmatrix} 0 & 0 \\ \mathbf{u} \otimes T \otimes \phi \otimes (\mathbf{s}\mathbf{e}_1^T(n_s+1)) & U \otimes T \otimes (\mathbf{s}\boldsymbol{\beta}) \end{bmatrix},$$

$$A_1 = \begin{bmatrix} T \otimes V \otimes I_{n_s+1} & T \otimes (\mathbf{v}\gamma) \otimes Q^* \\ \mathbf{u} \otimes [(\mathbf{t}\boldsymbol{\alpha}) \otimes \phi \otimes (\mathbf{s}\mathbf{e}_1^T(n_s+1)) + T \otimes \phi \otimes S^*] & U \otimes [T \otimes S + (\mathbf{t}\boldsymbol{\alpha}) \otimes (\mathbf{s}\boldsymbol{\beta})] \end{bmatrix},$$

$$B = T \otimes (V + \mathbf{v}\gamma), \; C = \begin{bmatrix} (\mathbf{t}\boldsymbol{\alpha}) \otimes V & (\mathbf{t}\boldsymbol{\alpha}) \otimes (\mathbf{v}\gamma) \otimes \boldsymbol{\beta} \end{bmatrix},$$

$$E = \begin{bmatrix} 0 \\ (\mathbf{u} + U\mathbf{e}^T) \otimes T \otimes \phi \otimes \mathbf{s} \end{bmatrix},$$

where I_j is an identity matrix of dimension j, $\mathbf{e}_j$ is the jth column of an identity matrix and $\mathbf{e}_j^T$ is its transpose, $S^* = [\mathbf{0} \;\; S]$ an $(n_s+1) \times n_s$ matrix, $Q^* = \begin{bmatrix} \boldsymbol{\beta} \\ Q \end{bmatrix}$, an $n_s \times (n_s+1)$ matrix. The dimensions of the zeros are left out to avoid complex notations, nevertheless they are straightforward to figure out.

If we allow the system to be a single vacation the only thing that is different are the boundary block matrices B, C and E and they are given as

$$B = \begin{bmatrix} T \otimes V & T \otimes (\mathbf{v}\gamma) \\ \mathbf{u} \otimes T \otimes \phi & U \otimes T \end{bmatrix},$$

$$C = \begin{bmatrix} (\mathbf{t}\boldsymbol{\alpha}) \otimes V \otimes \mathbf{e}_1^T(n_s+1) & \gamma \otimes (\mathbf{t}\boldsymbol{\alpha}) \times \mathbf{v} \otimes \boldsymbol{\beta} \\ \mathbf{u} \otimes (\mathbf{t}\boldsymbol{\alpha}) \otimes \phi \otimes \mathbf{e}_1^T(n_s+1) & U \otimes (\mathbf{t}\boldsymbol{\alpha}) \otimes \boldsymbol{\beta} \end{bmatrix},$$

$$E = \begin{bmatrix} 0 & 0 \\ \mathbf{u} \otimes T \otimes \phi \otimes \mathbf{s} & U \otimes T \otimes \mathbf{s} \end{bmatrix}.$$

We see that this problem is nicely set up as QBD and the results from QBD can be used to analyze this system. We first obtain the associated R matrix and then proceed as discussed in Chapter 2. However, in obtaining the R matrix here we notice that there is no obvious simple structure that can be exploited, hence it is more efficient to use the quadratically converging algorithms such as the Logarithmic Reduction [65] or the Cyclic Reduction [76].

It is immediately clear that if we let $\gamma = [1, 0, 0, \cdots, 0]$ and $U = \begin{bmatrix} \mathbf{0} & 0 \\ I & \mathbf{0}^T \end{bmatrix}$, then we have the time limited MAP/PH/1 vacation system presented in the last section. In addition if we consider a MAP/PH/1 queue in which the server may breakdown during service, go for repairs and resume service later, then we see that the MAP/PH/1 vacation queue with random vacation is actually the same model. The duration of a server's visit to a queue is simply the up time of the server and it's vacation period is the repair period. Hence the MAP/PH/1 vacation queue with random server visits is a very versatile queueing model.

5.3 Priority Queues

Most of the queue disciplines presented so far are the first-come-first-served systems. As pointed out earlier in the introductory part of single node systems other types of queue disciplines include the priority types, last-come-first-served, service in a random order, and several variants of these. In this section we present two types of priority service disciplines – the preemptive and non-preemptive types. We focus mainly on cases with just two classes of packets. The results for such a system can be used to approximate multiple classes systems, if needed. For the two classes presented in this section we assume there are two sources sending packets; one source sending one class type and the other sending another class type. The result is that we may have, in one time slot, no packet arrivals, just one packet arrival of either class, or two packet arrivals with one of each class. The other type of situation with no more than one packet arrival during a time slot is a special case of the model presented here. We study the priority queues with geometric arrivals and geometric services in detail.

5.3.1 Geo/Geo/1 Preemptive

Consider a system with two classes of packets that arrive according to the Bernoulli process with parameters a_k, $k = 1,2$, where $k = 1$ refers to a high priority (HP) packet and $k = 2$ refers to a lower priority (LP) packet. We consider the preemptive-resume discipline. The service discipline is as follows. No LP packet can start receiving service if there is an HP packet in the system and if an LP packet is receiving service (when there is no HP packet in the system), this LP packet service will be interrupted if an HP packet arrives before the LP's service is completed. An interrupted service is resumed where it was stopped, even though this information is irrelevant since service is geometric. Each packet k requires geometric service with parameter b_k. Since we are dealing in discrete time then during a time slot there is the possibility of having up to two packets of each type arriving. Hence we define $a_{i,j}$ as the probability of arrivals of i type HP packet and j arrivals of type LP packet arriving, $i = 0,1; j = 0,1$. Hence we have

$$a_{0,0} = (1-a_1)(1-a_2), \tag{5.78}$$

$$a_{0,1} = (1-a_1)a_2, \tag{5.79}$$

$$a_{1,0} = a_1(1-a_2), \tag{5.80}$$

$$a_{1,1} = a_1 a_2. \tag{5.81}$$

We consider the preemptive priority system, and assume the system is stable. It is simple to show that provided that $\frac{a_1}{b_1} + \frac{a_2}{b_2} < 1$ the system will be stable. For a stable system let $x_{i,j}$ be the probability that we find i of type HP and j of type LP packets in the system, including the one receiving service (if there is one receiving service). We can write the steady state equations as follows:

$$x_{0,0} = x_{0,0}a_{0,0} + x_{1,0}a_{0,0}b_1 + x_{0,1}a_{0,0}b_2, \tag{5.82}$$

$$x_{1,0} = x_{0,0}a_{1,0} + x_{1,0}a_{1,0}b_1 + x_{0,1}a_{1,0}b_2 + x_{1,0}a_{0,0}(1-b_1), \tag{5.83}$$

$$x_{i,0} = x_{i+1,0}a_{0,0}b_1 + x_{i-1,0}a_{1,0}(1-b_1) + x_{i,0}a_{0,0}(1-b_1) + x_{i,0}a_{1,0}b_1, \; i \geq 2, \tag{5.84}$$

$$x_{0,1} = x_{0,2}a_{0,0}b_2 + x_{0,0}a_{0,1} + x_{0,1}a_{0,0}(1-b_2) + x_{0,1}a_{0,1}b_2$$
$$+ x_{1,1}a_{0,0}b_1 + x_{1,0}a_{0,1}b_1, \tag{5.85}$$

$$x_{0,j} = x_{0,j+1}a_{0,0}b_2 + x_{0,j-1}a_{0,1}(1-b_2) + x_{0,j}a_{0,0}(1-b_2) + x_{0,j}a_{0,1}b_2$$
$$+ x_{1,j}a_{0,0}b_1 + x_{1,j-1}a_{0,1}b_1, \; j \geq 2, \tag{5.86}$$

$$x_{1,1} = x_{0,0}a_{1,1} + x_{2,0}a_{0,1}b_1 + x_{2,1}a_{0,0}b_1 + x_{0,1}a_{1,0}(1-b_2)$$
$$+ x_{1,0}a_{0,1}(1-b_1) + x_{1,1}a_{0,0}(1-b_1) + x_{1,1}a_{1,0}b_1, \qquad (5.87)$$

$$x_{i,j} = x_{i-1,j-1}a_{1,1}(1-b_1) + x_{i+1,j-1}a_{0,1}b_1 + x_{i+1,j}a_{0,0}b_1 + x_{i-1,j}a_{1,0}(1-b_1)$$
$$+ x_{i,j-1}a_{0,1}(1-b_1) + x_{i,j}a_{0,0}(1-b_1) + x_{i,j}a_{1,0}b_1, \; i > 1, j \geq 1, \; i \geq 1, j > 1. \qquad (5.88)$$

Even though we can proceed to analyze this system of equations using the traditional approaches, i.e. the algebraic approach and/or the $z-$transform approach, the effort involved is disproportionately high compared to the results to be obtained. However, we can use the matrix-geometric approach with less effort and obtain very useful results. We proceed with the matrix-geometric method as follows. First let us write

$$\boldsymbol{x}_i = [x_{i,0}, \, x_{i,1}, \, x_{i,2}, \, \cdots], \; \boldsymbol{x} = [\boldsymbol{x}_0, \, \boldsymbol{x}_1, \, \boldsymbol{x}_2, \, \cdots].$$

Consider the following state space $\Delta = \{(i_1, i_2), i_1 \geq 0, i_2 \geq 0\}$. This state space describes the DTMC that represents this preemptive priority system and the associated transition matrix is of the form

$$P = \begin{bmatrix} B_{0,0} & B_{0,1} & & & & \\ B_{1,0} & B_{1,1} & A_0 & & & \\ & A_2 & A_1 & A_0 & & \\ & & A_2 & A_1 & A_0 & \\ & & & A_2 & A_1 & A_0 \\ & & & & \ddots & \ddots & \ddots \end{bmatrix},$$

where

$$B_{0,0} = \begin{bmatrix} B_{00}^{00} & B_{00}^{01} & & & & \\ B_{00}^{10} & B_{00}^{1} & B_{00}^{0} & & & \\ & B_{00}^{2} & B_{00}^{1} & B_{00}^{0} & & \\ & & B_{00}^{2} & B_{00}^{1} & B_{00}^{0} & \\ & & & B_{00}^{2} & B_{00}^{1} & B_{00}^{0} \\ & & & & \ddots & \ddots & \ddots \end{bmatrix},$$

with

$$B_{00}^{00} = a_{0,0}, \; B_{00}^{01} = a_{0,1}, \; B_{00}^{10} = a_{0,0}b_2, \; B_{00}^{0} = a_{0,1}(1-b_2),$$
$$B_{00}^{1} = a_{0,0}(1-b_2) + a_{0,1}b_2, \; B_{00}^{2} = a_{0,0}b_2,$$

$$B_{0,1} = \begin{bmatrix} B_{01}^{00} & B_{01}^{01} & & & & \\ B_{01}^{2} & B_{01}^{1} & B_{01}^{0} & & & \\ & B_{01}^{2} & B_{01}^{1} & B_{01}^{0} & & \\ & & B_{01}^{2} & B_{01}^{1} & B_{01}^{0} & \\ & & & B_{01}^{2} & B_{01}^{1} & B_{01}^{0} \\ & & & & \ddots & \ddots & \ddots \end{bmatrix},$$

with

$$B_{01}^{00} = a_{1,0},\ B_{01}^{01} = a_{1,1},\ B_{01}^{2} = a_{1,0}b_2,\ B_{01}^{0} = a_{1,1}(1-b_2),$$
$$B_{01}^{1} = a_{1,0}(1-b_2) + a_{1,1}b_2,$$

$$B_{1,0} = \begin{bmatrix} B_{10}^{00} & B_{10}^{01} & & & & & \\ & B_{10}^{1} & B_{10}^{0} & & & & \\ & & B_{10}^{1} & B_{10}^{0} & & & \\ & & & B_{10}^{1} & B_{10}^{0} & & \\ & & & & B_{10}^{1} & B_{10}^{0} & \\ & & & & & \ddots & \ddots \end{bmatrix},$$

with

$$B_{10}^{00} = a_{0,0}b_1,\ B_{10}^{01} = a_{0,1}b_1,\ B_{10}^{0} = a_{0,1}b_1,\ B_{10}^{1} = a_{0,0}b_1,$$

$$B_{1,1} = \begin{bmatrix} B_{11}^{1} & B_{11}^{0} & & & & & \\ & B_{11}^{1} & B_{11}^{0} & & & & \\ & & B_{11}^{1} & B_{11}^{0} & & & \\ & & & B_{11}^{1} & B_{11}^{0} & & \\ & & & & B_{11}^{1} & B_{11}^{0} & \\ & & & & & \ddots & \ddots \end{bmatrix},$$

with

$$B_{11}^{1} = a_{0,0}b_1 + a_{1,0}b_1,\ B_{11}^{0} = a_{1,1}b_1 + a_{0,1}(1-b_1),$$

$$A_0 = \begin{bmatrix} A_0^1 & A_0^0 & & \\ & A_0^1 & A_0^0 & \\ & & \ddots & \ddots \end{bmatrix},\ A_0^1 = a_{1,0}(1-b_1),\ A_0^0 = a_{1,1}(1-b_1),$$

$$A_2 = \begin{bmatrix} A_2^1 & A_2^0 & & \\ & A_2^1 & A_2^0 & \\ & & \ddots & \ddots \end{bmatrix},\ A_2^1 = a_{0,0}b_1,\ A_2^0 = a_{0,1}b_1,$$

$$A_1 = \begin{bmatrix} A_1^1 & A_1^0 & & \\ & A_1^1 & A_1^0 & \\ & & \ddots & \ddots \end{bmatrix},\ A_1^1 = a_{0,0}(1-b_1) + a_{1,0}b_1,\ A_1^0 = a_{0,1}(1-b_1) + a_{1,1}b_1.$$

5.3.1.1 Stationary Distribution

If the system is stable we know that there is a relationship between the vector $\boldsymbol{x}$ and the matrix P of the form

$$\boldsymbol{x} = \boldsymbol{x}P,\ \ \boldsymbol{x}\mathbf{1} = 1,$$

and there exists a matrix R for which

$$\boldsymbol{x}_{i+1} = \boldsymbol{x}_i R,$$

where R is the minimal nonnegative solution to the matrix quadratic equation

$$R = A_0 + RA_1 + R^2 A_2.$$

It was shown by Alfa [13] and earlier by Miller [77] for the continuous time case that the matrix R has the following structure

$$R = \begin{bmatrix} r_0 & r_1 & r_2 & r_3 & \cdots \\ & r_0 & r_1 & r_2 & \cdots \\ & & r_0 & r_1 & \cdots \\ & & & r_0 & \ddots \\ & & & & \ddots \end{bmatrix}. \tag{5.89}$$

The arguments leading to this result are as follow. The matrix R is the expected number of visits to the HP state $i+1$ before first returning to the HP state i, given that the process started from state i. By this definition of the matrix R it is straightforward to see how the upper triangular pattern arises, i.e. because the LP number in the system can not reduce when there is an HP in the system. Since the number of LP arrivals is independent of the state of the number of the HP in the system, the rows of R will be repeating. This strucure makes it easier to study the matrix R.

Solving for R involves a recursive scheme as follows:

Let

$$f = 1 - a_{0,0}b_1 - a_{1,0}(1-b_1),$$

then we have

$$r_0 = \frac{f - \sqrt{(f^2 - 4a_{1,0}b_1a_{0,0}(1-b_1))}}{2a_{0,0}(1-b_1)}. \tag{5.90}$$

The term r_1 is obtained as

$$r_1 = \frac{a_{1,1}b_1 + r_0(a_{0,1}b_1 + a_{1,1}(1-b_1)) + (r_0)^2 a_{0,1}(1-b_1)}{f - 2r_0a_{0,0}(1-b_1)}, \tag{5.91}$$

and the remaining $r_j,\ j \geq 2$ are obtained as

$$r_j = \frac{ur_{j-1} + v\sum_{k=0}^{j-1} r_{j-k-1}r_k + w\sum_{k=1}^{j-1} r_{j-1}r_1}{f - 2r_0a_{0,0}(1-b_1)}, \tag{5.92}$$

where $u = a_{0,1}b_1 + a_{1,1}(1-b_1)),\ v = a_{0,1}(1-b_1),\ w = a_{0,0}(1-b_1)$. It was pointed out in Alfa [13] that $r_{j+1} \leq r_j,\ j \geq 1$, hence we can truncate R at an appropriate point of accuracy desired.

We need to obtain the boundary results $[\boldsymbol{x}_0,\ \boldsymbol{x}_1]$ and this is carried out using the boundary equation that

$$[\boldsymbol{x}_0,\ \boldsymbol{x}_1] = [\boldsymbol{x}_0,\ \boldsymbol{x}_1]B[R],\ [\boldsymbol{x}_0,\ \boldsymbol{x}_1]\mathbf{1} = 1,$$

where

$$B[R] = \begin{bmatrix} B_{00} & B_{01} \\ B_{10} & B_{11} + RA_2 \end{bmatrix}.$$

This can be solved using censored Markov chain idea to obtain $\boldsymbol{x}_0$ and after routine algebraic operations we have

$$\boldsymbol{x}_1 = \boldsymbol{x}_0 B_{01}(I - B_{11} - RA_1)^{-1}.$$

Note that for most practical situations the matrix $(I - B_{11} - RA_1)$ is non-singular. The vector $\boldsymbol{x}_0$ is normalized as

$$\boldsymbol{x}_0\mathbf{1} + \boldsymbol{x}_0[B_{01}(I - B_{11} - RA_1)^{-1}(I-R)^{-1}]\mathbf{1} = 1.$$

Both the matrices $(I - B_{11} - RA_1)$ and $I - R$ have the following triangular structure

$$U = \begin{bmatrix} u_0 & u_1 & u_2 & u_3 & \cdots \\ & u_0 & u_1 & u_2 & \cdots \\ & & u_0 & u_1 & \cdots \\ & & & u_0 & \ddots \\ & & & & \ddots \end{bmatrix}.$$

Their inverses also have the same structure and can be written as

$$V = \begin{bmatrix} v_0 & v_1 & v_2 & v_3 & \cdots \\ & v_0 & v_1 & v_2 & \cdots \\ & & v_0 & v_1 & \cdots \\ & & & v_0 & \ddots \\ & & & & \ddots \end{bmatrix},$$

where

$$v_0 = u_0^{-1}, \text{ and } v_n = -u_0^{-1}\sum_{j=0}^{n-1} u_{n-j}v_j,\ n \geq 1.$$

Let $y_i(k)$ be the probability that there are i type k packets in the system, $k = 1,2,\ i \geq 0$. We have

$$y_i(1) = \boldsymbol{x}_i\mathbf{1}, \text{ and } y_i(2) = x_{0,i} + \boldsymbol{x}_1(I-R)^{-1}\mathbf{e}_{i+1},$$

where $\mathbf{e}_j$ is the jth column of an infinite identity matrix.

The mean number of type 1, i.e. HP, in the system $E[X(1)]$ is given as

$$E[X(1)] = \sum_{j=1}^{\infty} j\boldsymbol{x}_j\mathbf{1} = \boldsymbol{x}_1(I-R)^{-1}\mathbf{1}, \tag{5.93}$$

and for type 2, i.e. LP, in the system $E[X(2)]$ is given as

$$E[X(2)] = \sum_{i=0}^{\infty}\sum_{j=1}^{\infty} jx_{i,j} = \boldsymbol{x}_0\phi + \boldsymbol{x}_1(I-R)^{-1}\phi, \tag{5.94}$$

where $\phi = [1,\ 2,\ 3,\ \cdots]^T$.

The waiting time for class 1, is simply obtained in the same manner as the waiting time in the Geo/Geo/1 system since a class 2 packet has no impact on the waiting time of type 1 packet. Let $W^q(1)$ be the waiting time in the queue for a type 1 packet and let $w_i^q(1) = Pr\{W^q(1) = i\}$ we have

$$w_0^q(1) = \boldsymbol{x}_0\mathbf{1}, \tag{5.95}$$

and

$$w_i^q(1) = \sum_{v=1}^{i} \boldsymbol{x}_v\mathbf{1}\binom{i-1}{v-1} b_1^v(1-b_1)^{i-v},\ \ i \geq 1. \tag{5.96}$$

Let $W(1)$ be the waiting time in the system (including service time) for class 1 packet, with $w_i(1) = Pr\{W(1) = i\}$, then we have

$$w_i(1) = \sum_{j=0}^{i-1} w_j^q(1)(1-b_1)^{i-j-1}b_1,\ \ i \geq 1. \tag{5.97}$$

The expected waiting time in the system for class 1 packets $E[W(1)]$ is obtained using the Little's Law as

$$E[W(1)] = a_1^{-1}E[X(1)], \tag{5.98}$$

and $E[W^q(1)]$ can be similarly obtained by first computing the expected number in the queue for the class 1 packets as

$$E[X_q(1)] = \sum_{j=1}^{\infty}(j-1)\boldsymbol{x}_j\mathbf{1} = \boldsymbol{x}_1R(I-R)^{-2}, \tag{5.99}$$

hence

$$E[W^q(1)] = a_1^{-1}E[X_q(1)]. \tag{5.100}$$

The waiting time in the queue for class 2 packets is a bit more involved to obtain. Keep in mind that even when a type 2 packet is receiving service it can still be preempted if a class 1 arrives during the service. Hence we can only determine a class 2 packet's waiting until it becomes the leading waiting packet in the queue for the first time, i.e. no class 1 or 2 ahead of it waiting. However we have to keep in mind that it could become the leading class 2 packet but with class 1 packets that arrive during its wait in the queue becoming the leading packets in the system. We can also study its waiting time in the system until it leaves the system. The later is a function of the former. We present an absorbing DTMC for studying the waiting time of a class 2 packet until it becomes the leading packet in the system for the first time.

Consider an absorbing DTMC with the following state space $\Delta = \{(i,j)\}, i \geq 0, j \geq 0$, with state $(0,0)$ as the absorbing state, where the first element refers to the number of HP packets in the system and the second element refers to the number of LP packets ahead of the target LP packet. The transition matrix of the absorbing DTMC $\tilde{P}$ is given as

$$\tilde{P} = \begin{bmatrix} \tilde{B}_{0,0} & \tilde{B}_{0,1} & & & & \\ \tilde{B}_{1,0} & \tilde{B}_{1,1} & \tilde{A}_0 & & & \\ & \tilde{A}_2 & \tilde{A}_1 & \tilde{A}_0 & & \\ & & \tilde{A}_2 & \tilde{A}_1 & \tilde{A}_0 & \\ & & & \tilde{A}_2 & \tilde{A}_1 & \tilde{A}_0 \\ & & & & \ddots & \ddots & \ddots \end{bmatrix},$$

where

$$\tilde{B}_{0,0} = \begin{bmatrix} 1 & & & & & \\ \tilde{B}_{00}^2 & \tilde{B}_{00}^1 & & & & \\ & \tilde{B}_{00}^2 & \tilde{B}_{00}^1 & & & \\ & & \tilde{B}_{00}^2 & \tilde{B}_{00}^1 & & \\ & & & \tilde{B}_{00}^2 & \tilde{B}_{00}^1 & \\ & & & & \ddots & \ddots & \ddots \end{bmatrix},$$

with

$$\tilde{B}_{00}^1 = (1-a_1)(1-b_2),\; \tilde{B}_{00}^2 = (1-a_1)b_2,$$

$$\tilde{B}_{0,1} = \begin{bmatrix} 0 & 0 & & & & \\ \tilde{B}_{01}^2 & \tilde{B}_{01}^1 & & & & \\ & \tilde{B}_{01}^2 & \tilde{B}_{01}^1 & & & \\ & & \tilde{B}_{01}^2 & \tilde{B}_{01}^1 & & \\ & & & \tilde{B}_{01}^2 & \tilde{B}_{01}^1 & \\ & & & & \ddots & \ddots & \ddots \end{bmatrix},$$

with

$$\tilde{B}_{01}^2 = a_1 b_2,\; \tilde{B}_{01}^1 = a_1(1-b_2),$$

$$\tilde{B}_{1,0} = \begin{bmatrix} \tilde{B}_{10}^{00} & & & & & \\ & \tilde{B}_{10}^1 & & & & \\ & & \tilde{B}_{10}^1 & & & \\ & & & \tilde{B}_{10}^1 & & \\ & & & & \tilde{B}_{10}^1 & \\ & & & & & \ddots & \ddots \end{bmatrix},$$

with

$$\tilde{B}_{10}^{00} = (1-a_1)b_1,\; \tilde{B}_{10}^1 = (1-a_1)b_1,$$

$$\tilde{A}_0 = Ia_1(1-b_1),\; \tilde{A}_1 = \tilde{B}_{11} = I((1-a_1)(1-b_1)+a_1 b_1),\; \tilde{A}_2 = I(1-a_1)b_1.$$

Now we define $\mathbf{z}^{(0)} = [\mathbf{z}_0^{(0)}, \mathbf{z}_1^{(0)}, \mathbf{z}_2^{(0)}, \cdots]$ with $\mathbf{z}_i^{(0)} = [z_{i,0}^{(0)}, z_{i,1}^{(0)}, z_{i,2}^{(0)}, \cdots]$, $i \geq 0$, and let $\mathbf{z}^{(0)} = \boldsymbol{x}$. Define $W^*(2)$ as the waiting time in the queue for a type 2 packet until it becomes the leading packet waiting in the system for the first time, and $W_j^* = Pr\{W^*(2) \leq j\}$ then we have

$$W_i^* = z_{0,0}^{(i)}, \; i \geq 1, \text{ where } \mathbf{z}^{(i)} = \mathbf{z}^{(i-1)}\tilde{P}, \tag{5.101}$$

keeping in mind that

$$W_0^* = z_{0,0}^{(0)} = x_{0,0}. \tag{5.102}$$

5.3.2 Geo/Geo/1 Non-preemptive

In a non-preemptive system no LP packet's service is going to be started if there is an HP packet in the system. However if an LP packet's service is started when there is no HP packet in the system, this LP packet's service will not be interrupted, i.e. no preemption.

Now we consider a system with two classes of packets that arrive according to the Bernoulli process with parameters a_k, $k = 1,2$, where $k = 1$ refers to a high priority (HP) packet and $k = 2$ refers to a lower priority (LP) packet. Each packet k requires geometric service with parameter b_k. Since we are dealing in discrete time then during a time slot there is the possibility of having up to two packets, one of each class type, arriving. Hence we define $a_{i,j}$ as the probability of a i type HP packet and j of type LP packet arriving, $(i,j) = 0,1$. Hence we have

$$a_{0,0} = (1-a_1)(1-a_2), \; a_{0,1} = (1-a_1)a_2, \; a_{1,0} = a_1(1-a_2), \; a_{1,1} = a_1 a_2.$$

We assume the system is stable. It is simple to show that provided $\frac{a_1}{b_1} + \frac{a_2}{b_2} < 1$ the system will be stable. For a stable system let $x_{i,j,k}$ be the probability that we find i number of type HP and j number of type LP packets in the system, including the one receiving system (if there is one receiving service) and the one receiving service is type k. Here we notice that we have to keep track of which type of packet is receiving service since it is a non-preemptive system. It is immediately clear that the steady state equations for this non-preemptive system is going to be more involved. We skip this step and use the matrix-geometric method directly. First let us write

$$\boldsymbol{x}_i = [\boldsymbol{x}_{i,0}, \boldsymbol{x}_{i,1}, \boldsymbol{x}_{i,2}, \boldsymbol{x}_{i,3}, \cdots], \; \boldsymbol{x} = [\boldsymbol{x}_0, \boldsymbol{x}_1, \boldsymbol{x}_2, \cdots],$$

and write $\boldsymbol{x}_{i,j} = [x_{i,j,1}, x_{i,j+1,2}], i \geq 1, j \geq 0$. The state space for this system can be written as

$$\Delta = \Delta_1 \cup \Delta_2 \cup \Delta_3,$$

where

- $\Delta_1 = \{(0,0)\}$,

- $\Delta_2 = \{(0,j),\ j \geq 1,\}$,
- $\Delta_3 = \{(i,j+k-1,k),\ i \geq 1, j \geq 0, k = 1,2.\}$

The transition matrix P associated with this DTMC is given as

$$P = \begin{bmatrix} B_{0,0} & B_{0,1} & & & & \\ B_{1,0} & A_1 & A_0 & & & \\ & A_2 & A_1 & A_0 & & \\ & & A_2 & A_1 & A_0 & \\ & & & A_2 & A_1 & A_0 \\ & & & & \ddots & \ddots & \ddots \end{bmatrix},$$

where

$$B_{0,0} = \begin{bmatrix} B_{00}^{00} & B_{00}^{01} & & & & \\ B_{00}^{10} & B_{00}^{1} & B_{00}^{0} & & & \\ & B_{00}^{2} & B_{00}^{1} & B_{00}^{0} & & \\ & & B_{00}^{2} & B_{00}^{1} & B_{00}^{0} & \\ & & & B_{00}^{2} & B_{00}^{1} & B_{00}^{0} \\ & & & & \ddots & \ddots & \ddots \end{bmatrix},$$

with

$$B_{00}^{00} = a_{0,0},\ B_{00}^{01} = a_{0,1},\ B_{00}^{10} = a_{0,0}b_2,\ B_{00}^{0} = a_{0,1}(1-b_2),$$

$$B_{00}^{1} = a_{0,0}(1-b_2) + a_{0,1}b_2,\ B_{00}^{2} = a_{0,0}b_2,$$

$$B_{0,1} = \begin{bmatrix} B_{01}^{00} & B_{01}^{01} & & & \\ B_{01}^{2} & B_{01}^{1} & & & \\ & B_{01}^{2} & B_{01}^{1} & & \\ & & B_{01}^{2} & B_{01}^{1} & \\ & & & B_{01}^{2} & B_{01}^{1} \\ & & & & \ddots & \ddots \end{bmatrix},$$

with

$$B_{01}^{00} = [a_{1,0},\ 0],\ B_{01}^{01} = [a_{1,1},\ 0],\ B_{01}^{2} = [a_{1,0}b_2,\ a_{1,0}(1-b_2)],$$

$$B_{01}^{1} = [a_{1,1}b_2,\ a_{1,1}(1-b_2)],$$

$$B_{1,0} = \begin{bmatrix} B_{10}^{00} & B_{10}^{0} & & & & \\ & B_{10}^{1} & B_{10}^{0} & & & \\ & & B_{10}^{1} & B_{10}^{0} & & \\ & & & B_{10}^{1} & B_{10}^{0} & \\ & & & & B_{10}^{1} & B_{10}^{0} \\ & & & & & \ddots & \ddots \end{bmatrix},$$

with

$$B_{10}^{00} = \begin{bmatrix} a_{0,0}b_1 \\ 0 \end{bmatrix},\ B_{10}^{0} = \begin{bmatrix} a_{0,1}b_1 \\ 0 \end{bmatrix},\ B_{10}^{1} = \begin{bmatrix} a_{0,0}b_1 \\ 0 \end{bmatrix},$$

$$A_0 = \begin{bmatrix} A_0^1 & A_0^0 & & \\ & A_0^1 & A_0^0 & \\ & & \ddots & \ddots \end{bmatrix}, \quad A_0^1 = \begin{bmatrix} a_{1,0}(1-b_1) & 0 \\ a_{1,0}b_2 & a_{1,0}(1-b_2) \end{bmatrix},$$

$$A_0^0 = \begin{bmatrix} a_{1,1}(1-b_1) & 0 \\ a_{1,1}b_2 & a_{1,1}(1-b_2) \end{bmatrix},$$

$$A_2 = \begin{bmatrix} A_2^1 & A_2^0 & & \\ & A_2^1 & A_2^0 & \\ & & \ddots & \ddots \end{bmatrix}, \quad A_2^1 = \begin{bmatrix} a_{0,0}b_1 & 0 \\ 0 & 0 \end{bmatrix}, \quad A_2^0 = \begin{bmatrix} a_{0,1}b_1 & 0 \\ 0 & 0 \end{bmatrix},$$

$$A_1 = \begin{bmatrix} A_1^1 & A_1^0 & & \\ & A_1^1 & A_1^0 & \\ & & \ddots & \ddots \end{bmatrix}, \quad A_1^0 = \begin{bmatrix} a_{0,1}(1-b_1)+a_{1,1}b_1 & 0 \\ a_{0,1}b_2 & a_{0,1}(1-b_2) \end{bmatrix},$$

$$A_1^1 = \begin{bmatrix} a_{0,0}(1-b_1)+a_{1,0}b_1 & 0 \\ a_{0,0}b_2 & a_{0,0}(1-b_2) \end{bmatrix}.$$

5.3.2.1 Stationary Distribution

If the system is stable we know that there is a relationship between the vector $\boldsymbol{x}$ and the matrix P of the form

$$\boldsymbol{x} = \boldsymbol{x}P, \ \ \boldsymbol{x}\mathbf{1} = 1,$$

and there exists a matrix R for which

$$\boldsymbol{x}_{i+1} = \boldsymbol{x}_i R,$$

where R is the minimal nonnegative solution to the matrix quadratic equation

$$R = A_0 + RA_1 + R^2 A_2.$$

It was shown by Alfa [13] and earlier by Miller [77] for the continuous time case that the matrix R has the following structure

$$R = \begin{bmatrix} R_0 & R_1 & R_2 & R_3 & \cdots \\ & R_0 & R_1 & R_2 & \cdots \\ & & R_0 & R_1 & \cdots \\ & & & R_0 & \ddots \\ & & & & \ddots \end{bmatrix}. \tag{5.103}$$

The arguments leading to this result are as follow. The matrix R is the expected number of visits to the HP state $i+1$ before first returning to the HP state i, given that the process started from state i. By this definition of the matrix R it is straightforward to see how the upper triangular pattern arises, i.e. because the LP number in the system cannot reduce when there is an HP in the system. Since the number of LP arrivals is independent of the state of the number of the HP in the system, the rows of R will be repeating. This structure makes it easier to study the matrix R. However, the R_i, $i \geq 0$ also has an additional structure. We have

$$R_i = \begin{bmatrix} r_i(1,1) & 0 \\ r_i(2,1) & r_i(2,2) \end{bmatrix}. \tag{5.104}$$

The matrix R can be computed to a suitable truncation level and then used to compute the stationary vector after the boundary equation has been solved. We have

$$\boldsymbol{x}_{i+1} = \boldsymbol{x}_i R, \; i \geq 1, \tag{5.105}$$

and specifically we have

$$x_{i+1,j,1} = \sum_{v=0}^{j} [x_{i,j-v,1} r_v(1,1) + x_{i,j-v+1,2} r_v(2,1)], \; i \geq 1, \; j \geq 0, \tag{5.106}$$

and

$$x_{i+1,j+1,2} = \sum_{v=0}^{j} x_{i,j-v+1,2} r_{j-v}(2,2), \; i \geq 1, \; j \geq 0. \tag{5.107}$$

5.3.3 MAP/PH/1 Preemptive

Consider a system with two classes of packets that arrive according to the Markovian arrival process represented by four matrices $D_{0,0}$, $D_{1,0}$, $D_{0,1}$, and $D_{1,1}$ all of dimension $n_t \times n_t$, where D_{i_1,i_2} refers to i_1 arrivals of type 1 (HP) and i_2 arrivals of type 2 (LP). For example, $D_{0,1}$ refers to no arrival of type 1 (the HP class) packet and 1 arrival of type 2 (the LP class) packet. The matrix $D = D_{0,0} + D_{1,0} + D_{0,1} + D_{1,1}$ is stochastic. Let us assume that it is irreducible and let its stationary distribution be $\boldsymbol{\pi} = \boldsymbol{\pi} D$, $\boldsymbol{\pi}\mathbf{1} = 1$, then $a_1 = \boldsymbol{\pi}(D_{1,0} + D_{1,1})\mathbf{1}$ and $a_2 = \boldsymbol{\pi}(D_{0,1} + D_{1,1})\mathbf{1}$, where a_k is the probability of a type k arrival in a time slot. Packets of type k class have PH service times with parameters $(\boldsymbol{\beta}_k, S_k)$ of order $n_s(k)$, $k = 1,2$. Let $(b_k)^{-1} = \boldsymbol{\beta}_k(I - S_k)^{-1}\mathbf{1}$. We consider the preemptive discipline. The service discipline is as follows. No LP packet can start receiving service if there is an HP packet in the system and if an LP packet is receiving service (when there is no HP packet in the system), this LP packet service will be interrupted if an HP packet arrives before its service is completed. An interrupted service in phase j_1 is resumed in phase j_2 with probability Q_{j_1,j_2} which is an element of the stochastic matrix Q. If for example, $Q = I$ then an interrupted service starts from the phase where it was stopped, i.e. preemptive resume whereas if $Q = \mathbf{1}\boldsymbol{\beta}_1$ then the service is started from beginning again, i.e. preemptive repeat. We know that a_k, and b_k are the mean arrival rates and service rates, respectively of class k packet. It is simple to show that provided that $\frac{a_1}{b_1} + \frac{a_2}{b_2} < 1$ the system will be stable. This system is easier to study using the matrix-analytic method.

Consider the following state space

$$\Delta = \{(0,0,\ell) \cup (0,i_2,\ell,j_2) \cup (i_1,i_2-1,\ell,j_1,j_2^{'}),\ \ell = 1,2,\cdots,n_t; i_1 \geq 1; i_2 \geq 1;$$

$$j_1 = 1,2,\cdots,n_s(1);\ j_2 = 1,2,\cdots,n_s(2); j_2^{'} = 1,2,\cdots,n_s(2)\ \}.$$

The elements of the state space are ℓ referring to the phase of arrival, i_v, $v = 1,2$, referring to the number of type v packets in the system, and j_v, $v = 1,2$ the phase of service of type v in service, and $j_2^{'}$ is the phase at which the type 2 packet's service resumes next time after preemption. The first tupple $(0,0,\ell)$ refers to an empty system, the second tupple $(0,i_2,\ell,j_2)$ refers to the system with only type 2 packets in the system and the third tupple (i_1,i_2,ℓ,j_1) refers to the case where there is at least one type 1 packet in the system, which is receiving service. This state space describes the DTMC that represents this preemptive priority system and the associated transition matrix is of the form

$$P = \begin{bmatrix} B_{0,0} & B_{0,1} & & & & \\ B_{1,0} & B_{1,1} & A_0 & & & \\ & A_2 & A_1 & A_0 & & \\ & & A_2 & A_1 & A_0 & \\ & & & A_2 & A_1 & A_0 \\ & & & & \ddots & \ddots & \ddots \end{bmatrix},$$

where

$$B_{0,0} = \begin{bmatrix} B_{00}^{00} & B_{00}^{01} & & & & \\ B_{00}^{10} & B_{00}^{1} & B_{00}^{0} & & & \\ & B_{00}^{2} & B_{00}^{1} & B_{00}^{0} & & \\ & & B_{00}^{2} & B_{00}^{1} & B_{00}^{0} & \\ & & & B_{00}^{2} & B_{00}^{1} & B_{00}^{0} \\ & & & & \ddots & \ddots & \ddots \end{bmatrix},$$

with

$$B_{00}^{00} = D_{0,0},\ B_{00}^{01} = D_{0,1} \otimes \boldsymbol{\beta}_2,\ B_{00}^{10} = D_{0,0}\mathbf{s}_2,\ B_{00}^{0} = D_{0,1} \otimes S_2,$$

$$B_{00}^{1} = D_{0,0} \otimes S_2 + D_{0,1} \otimes (\mathbf{s}_2\boldsymbol{\beta}_2),\ B_{00}^{2} = D_{0,0} \otimes (\mathbf{s}_2\boldsymbol{\beta}_2),$$

$$B_{0,1} = \begin{bmatrix} B_{01}^{00} & B_{01}^{01} & & & & \\ B_{01}^{2} & B_{01}^{1} & B_{01}^{0} & & & \\ & B_{01}^{2} & B_{01}^{1} & B_{01}^{0} & & \\ & & B_{01}^{2} & B_{01}^{1} & B_{01}^{0} & \\ & & & B_{01}^{2} & B_{01}^{1} & B_{01}^{0} \\ & & & & \ddots & \ddots & \ddots \end{bmatrix},$$

with

$$B_{01}^{00} = D_{1,0} \otimes \boldsymbol{\beta}_1,\ B_{01}^{01} = D_{1,1} \otimes \boldsymbol{\beta}_1,\ B_{01}^{2} = D_{1,0} \otimes (\mathbf{s}_2 \boldsymbol{\beta}_1),\ B_{01}^{0} = D_{1,1} \otimes (S_2 Q) \otimes \boldsymbol{\beta}_1,$$

$$B_{01}^{1} = D_{1,0} \otimes (S_2 Q) \otimes \boldsymbol{\beta}_1 + D_{1,1}(\mathbf{s}_2 \boldsymbol{\beta}_2) Q \otimes \boldsymbol{\beta}_1,$$

$$B_{1,0} = \begin{bmatrix} B_{10}^{00} & B_{10}^{01} & & & & & \\ & B_{10}^{1} & B_{10}^{0} & & & & \\ & & B_{10}^{1} & B_{10}^{0} & & & \\ & & & B_{10}^{1} & B_{10}^{0} & & \\ & & & & B_{10}^{1} & B_{10}^{0} & \\ & & & & & \ddots & \ddots \end{bmatrix},$$

with

$$B_{10}^{00} = D_{0,0} \otimes \mathbf{s}_1,\ B_{10}^{01} = D_{0,1} \otimes (\mathbf{s}_1 \boldsymbol{\beta}_1),\ B_{10}^{0} = D_{0,1} \otimes (\mathbf{s}_1 \boldsymbol{\beta}_2),\ B_{10}^{1} = D_{0,0} \otimes (\mathbf{s}_1 \boldsymbol{\beta}_2),$$

$$B_{1,1} = \begin{bmatrix} B_{11}^{1} & B_{11}^{0} & & & & & \\ & B_{11}^{1} & B_{11}^{0} & & & & \\ & & B_{11}^{1} & B_{11}^{0} & & & \\ & & & B_{11}^{1} & B_{11}^{0} & & \\ & & & & B_{11}^{1} & B_{11}^{0} & \\ & & & & & \ddots & \ddots \end{bmatrix},$$

with

$$B_{11}^{1} = (D_{0,0} \otimes S_1 + D_{1,0} \otimes (\mathbf{s}_1 \boldsymbol{\beta}_1)) \otimes Q,\ B_{11}^{0} = (D_{1,1} \otimes (\mathbf{s}_1 \boldsymbol{\beta}_1) + D_{0,1} \otimes S_1) \otimes Q,$$

$$A_0 = \begin{bmatrix} A_0^1 & A_0^0 & & \\ & A_0^1 & A_0^0 & \\ & & \ddots & \ddots \end{bmatrix},\ A_0^1 = D_{1,0} \otimes S_1 \otimes Q,\ A_0^0 = D_{1,1} \otimes S_1 \otimes Q,$$

$$A_2 = \begin{bmatrix} A_2^1 & A_2^0 & & \\ & A_2^1 & A_2^0 & \\ & & \ddots & \ddots \end{bmatrix},\ A_2^1 = D_{0,0} \otimes (\mathbf{s}_1 \boldsymbol{\beta}_1) \otimes Q,\ A_2^0 = D_{0,1} \otimes (\mathbf{s}_1 \boldsymbol{\beta}_1) \otimes Q,$$

$$A_1 = \begin{bmatrix} A_1^1 & A_1^0 & & \\ & A_1^1 & A_1^0 & \\ & & \ddots & \ddots \end{bmatrix},$$

$$A_1^1 = (D_{0,0} \otimes S_1 + D_{1,0}(\mathbf{s}_1 \boldsymbol{\beta}_1)) \otimes Q,\ A_1^0 = (D_{0,1} \otimes S_1 + D_{1,1}(\mathbf{s}_1 \boldsymbol{\beta}_1)) \otimes Q.$$

5.3.3.1 Stationary Distribution

If the system is stable we know that there is a relationship between the vector $\boldsymbol{x}$ and the matrix P of the form

$$\boldsymbol{x} = \boldsymbol{x}P, \;\; \boldsymbol{x}\mathbf{1} = 1,$$

and there exists a matrix R for which

$$\boldsymbol{x}_{i+1} = \boldsymbol{x}_i R,$$

where R is the minimal nonnegative solution to the matrix quadratic equation

$$R = A_0 + RA_1 + R^2 A_2,$$

and

$$\boldsymbol{x} = [\boldsymbol{x}_0, \; \boldsymbol{x}_1, \; \boldsymbol{x}_2, \; \cdots].$$

It was shown by Alfa [13] and earlier by Miller [77] for the continuous time case that the matrix R has the following structure

$$R = \begin{bmatrix} R_0 & R_1 & R_2 & R_3 & \cdots \\ & R_0 & R_1 & R_2 & \cdots \\ & & R_0 & R_1 & \cdots \\ & & & R_0 & \ddots \\ & & & & \ddots \end{bmatrix}. \tag{5.108}$$

The arguments leading to this result are as follow. The matrix R is the expected number of visits to the HP state $i+1$ before first returning to the HP state i, given that the process started from state i. By this definition of the matrix R it is straightforward to see how the upper triangular pattern arises, i.e. because the LP number in the system can not reduce when there is an HP in the system. Since the number of LP arrivals is independent of the state of the number of the HP in the system, the rows of R will be repeating. This structure makes it easier to study the matrix R.

Solving for R involves a recursive scheme as follows:
First we write out the R equations in block form as

$$R_0 = A_0^1 + R_0 A_1^1 + (R_0)^2 A_2^1, \tag{5.109}$$

$$R_1 = A_0^0 + R_0 A_1^0 + R_1 A_1^1 + (R_0 R_1 + R_1 R_0) A_2^1, \tag{5.110}$$

$$R_j = R_{j-1} A_1^0 + R_j A_1^1 + \sum_{v=0}^{j-1} R_{j-v-1} R_v A_2^0 + \sum_{v=0}^{j} R_{j-v} R_v A_2^1, \; j \geq 2. \tag{5.111}$$

Let us write

$$F_2 = A_2^1,$$

$$F_0 = I - A_1^1 - R_0 F_2,$$

and

$$F_{1,j} = \delta_{1,j} A_0^1 + R_{j-1} A_1^0 + \sum_{v=0}^{j-1} R_{j-v-1} R_v A_2^0 + (1 - \delta_{1,j}) \sum_{v=0}^{j} R_{j-v} R_v F_2, \; j \geq 1,$$

where, δ_{ij} is the kronecker delta writen as $\delta = \begin{cases} 1, & i=j, \\ 0, & \text{otherwise} \end{cases}$, then we have the solution to the matrix sequence $R_j,\ j=1,2,\cdots$ given as

- R_0 is the minimal non-negative solution to the matrix quadratic equation

$$R_0 = A_0^1 + R_0 A_1^1 + (R_0)^2 A_2^1, \tag{5.112}$$

  which can be solved for using any of the special techniques for the QBD such as the Logarithmic Reduction or the Cyclic Reduction, and
- $R_j,\ j \geq 1$ are given as the solutions to the matrix linear equations

$$(F_0^T \otimes I - F_2^T \otimes R_0) vecR_j = vecF_{1,j}^T, \tag{5.113}$$

  where for an $M \times N$ matrix $Y = [Y_{.1},\ Y_{.2},\ \cdots,\ Y_{.j},\ \cdots,\ Y_{.N}]$, where $Y_{.j}$ is the jth column vector of the matrix Y, we define a $1 \times NM$ column vector

$$vecY = \begin{bmatrix} Y_{.1} \\ Y_{.2} \\ \vdots \\ Y_{.j} \\ \cdots \\ Y_{.N} \end{bmatrix}.$$

It was pointed out in Alfa [13] that $(R_{k+1})_{i,j} \leq (R_k)_{i,j},\ k \geq 1$, hence we can truncate R at an appropriate point of desired accuracy.

We need to obtain the boundary results $[\boldsymbol{x}_0,\ \boldsymbol{x}_1]$ and this is carried out using the boundary equation that

$$[\boldsymbol{x}_0,\ \boldsymbol{x}_1] = [\boldsymbol{x}_0,\ \boldsymbol{x}_1] B[R],\ [\boldsymbol{x}_0,\ \boldsymbol{x}_1]\mathbf{1} = 1,$$

where

$$B[R] = \begin{bmatrix} B_{00} & B_{01} \\ B_{10} & B_{11} + RA_2 \end{bmatrix}.$$

This can be solved using censored Markov chain idea to obtain $\boldsymbol{x}_0$ and after routine algebraic operations we have

$$\boldsymbol{x}_1 = \boldsymbol{x}_0 B_{01} (I - B_{11} - RA_1)^{-1}.$$

Note that for most practical situations the matrix $(I - B_{11} - RA_1)$ is non-singular. The vector $\boldsymbol{x}_0$ is normalized as

$$\boldsymbol{x}_0 \mathbf{1} + \boldsymbol{x}_0 [B_{01} (I - B_{11} - RA_1)^{-1} (I - R)^{-1}] \mathbf{1} = 1.$$

Both the matrices $(I - B_{11} - RA_1)$ and $I - R$ have the following triangular structure

$$U = \begin{bmatrix} U_0 & U_1 & U_2 & U_3 & \cdots \\ & U_0 & U_1 & U_2 & \cdots \\ & & U_0 & U_1 & \cdots \\ & & & U_0 & \ddots \\ & & & & \ddots \end{bmatrix}.$$

Their inverses also have the same structure and can be written as

$$V = \begin{bmatrix} V_0 & V_1 & V_2 & V_3 & \cdots \\ & V_0 & V_1 & V_2 & \cdots \\ & & V_0 & V_1 & \cdots \\ & & & V_0 & \ddots \\ & & & & \ddots \end{bmatrix},$$

where

$$V_0 = U_0^{-1}, \text{ and } V_\ell = -U_0^{-1} \sum_{j=0}^{\ell-1} U_{\ell-j} V_j, \ \ell \geq 1.$$

Given $\boldsymbol{x}$ the major performance measures such as the distribution, and the mean, of the number of type k, $k = 1,2$ packets in the system is straightforward to obtain. The waiting times for type 1 is straightforward to obtain also. It is the same idea as used for obtaining the waiting time of the one class MAP/PH/1 or PH/PH/1 model. However, the waiting for the type 2 class is more involved. For details of how to obtain all these performance measures see Alfa, Liu and He [4]. The details will be omitted from this book.

5.3.4 MAP/PH/1 Non-preemptive

Consider a system with two classes of packets that arrive according to the Markovian arrival process represented by four matrices $D_{0,0}$, $D_{1,0}$, $D_{0,1}$, and $D_{1,1}$ all of dimension $n_t \times n_t$, where D_{i_1,i_2} refers to i_k arrivals of type k, $k = 1,2$. For example, $D_{0,1}$ refers to no arrival of type 1 (the HP class) packet and 1 arrival of type 2 (the LP class) packet. The matrix $D = D_{0,0} + D_{1,0} + D_{0,1} + D_{1,1}$ is stochastic. Let us assume that it is irreducible and let its stationary distribution be $\boldsymbol{\pi} = \boldsymbol{\pi} D$, $\boldsymbol{\pi}\mathbf{1} = 1$, then $a_1 = \boldsymbol{\pi}(D_{1,0} + D_{1,1})\mathbf{1}$ and $a_2 = \boldsymbol{\pi}(D_{0,1} + D_{1,1})\mathbf{1}$, where again, a_k is the probability of a type k arrival during a time slot. Packets of type k class have PH service times with parameters $(\boldsymbol{\beta}_k, S_k)$ of order $n_s(k)$, $k = 1,2$. Let $(b_k)^{-1} = \boldsymbol{\beta}_k(I - S_k)^{-1}\mathbf{1}$. We consider the nonpreemptive discipline. The service discipline is as follows. No LP packet can start receiving service if there is an HP packet in the system. However if an LP packet starts to receive service when there is no HP packet in the system, this LP packet service will continue until completion even if an HP packet arrives before the service is completed. We know that a_k, b_k are the mean arrival rates and service rates, respectively of class k packet. It is simple to show that provided that $\frac{a_1}{b_1} + \frac{a_2}{b_2} < 1$ the system will be stable. This system is easier to study using the matrix-analytic method.

The state space for this system can be written as

$$\Delta = \Delta_1 \cup \Delta_2 \cup \Delta_3,$$

where

- $\Delta_1 = \{(0,0,\ell),\ \ell = 1,2,\cdots,n_t\}$,
- $\Delta_2 = \{(0,i_2,\ell,j_2),\ i_2 \geq 1,\ \ell = 1,2,\cdots,n_t; j_2 = 1,2,\cdots,n_s(2)\}$,
- $\Delta_3 = \{(i_1,i_2+v-1,v,\ell,j_v),\ i_1 \geq 1; i_2 \geq 0;\ v = 1,2;$ $\ell = 1,2,\cdots,n_t;\ j_v = 1,2,\cdots,n_s(v).\}$

Throughout this state space definition we have, $i_v,\ v = 1,2$ referring to the number of type v packets in the system, ℓ to the phase of arrival, and k_v to the phase of type v packet in service. The first tupple $(0,0,\ell)$ refers to an empty system, the second $(0,i_2,\ell,j_2)$ to having no type 1 and at least one type 2 packets in the system, and the third tupple (i_1,i_2+v-1,v,ℓ,j_v) refers to having i_1 and i_2+v-1 type 1 and type 2 packets in the system ,with a type v in service whose service is in phase j_v.

The transition matrix P associated with this DTMC is given as

$$P = \begin{bmatrix} B_{0,0} & B_{0,1} & & & & \\ B_{1,0} & A_1 & A_0 & & & \\ & A_2 & A_1 & A_0 & & \\ & & A_2 & A_1 & A_0 & \\ & & & A_2 & A_1 & A_0 \\ & & & & \ddots & \ddots & \ddots \end{bmatrix},$$

where

$$B_{0,0} = \begin{bmatrix} B_{00}^{00} & B_{00}^{01} & & & & \\ B_{00}^{10} & B_{00}^{1} & B_{00}^{0} & & & \\ & B_{00}^{2} & B_{00}^{1} & B_{00}^{0} & & \\ & & B_{00}^{2} & B_{00}^{1} & B_{00}^{0} & \\ & & & B_{00}^{2} & B_{00}^{1} & B_{00}^{0} \\ & & & & \ddots & \ddots & \ddots \end{bmatrix},$$

with

$$B_{00}^{00} = D_{0,0},\ B_{00}^{01} = D_{0,1} \otimes \boldsymbol{\beta}_2,\ B_{00}^{10} = D_{0,0} \otimes \mathbf{s}_2,\ B_{00}^{0} = D_{0,1} \otimes S_2,$$

$$B_{00}^{1} = D_{0,0} \otimes S_2 + D_{0,1} \otimes (\mathbf{s}_2\boldsymbol{\beta}_2),\ B_{00}^{2} = D_{0,0} \otimes (\mathbf{s}_2\boldsymbol{\beta}_2),$$

$$B_{0,1} = \begin{bmatrix} B_{01}^{00} & B_{01}^{01} & & & & \\ B_{01}^{2} & B_{01}^{1} & & & & \\ & B_{01}^{2} & B_{01}^{1} & & & \\ & & B_{01}^{2} & B_{01}^{1} & & \\ & & & B_{01}^{2} & B_{01}^{1} & \\ & & & & \ddots & \ddots \end{bmatrix},$$

with

$$B_{01}^{00} = [D_{1,0} \otimes \boldsymbol{\beta}_1,\ 0],\ B_{01}^{01} = [D_{1,1} \otimes \boldsymbol{\beta}_1,\ 0],\ B_{01}^{2} = [D_{1,0} \otimes (\mathbf{s}_2\boldsymbol{\beta}_1),\ D_{1,0} \otimes S_2],$$

$$B_{01}^{1} = [D_{1,1} \otimes (\mathbf{s}_2\boldsymbol{\beta}_1),\ D_{1,1} \otimes S_2],$$

$$B_{1,0} = \begin{bmatrix} B_{10}^{00} & B_{10}^{0} & & & & \\ & B_{10}^{1} & B_{10}^{0} & & & \\ & & B_{10}^{1} & B_{10}^{0} & & \\ & & & B_{10}^{1} & B_{10}^{0} & \\ & & & & B_{10}^{1} & B_{10}^{0} \\ & & & & \ddots & \ddots \end{bmatrix},$$

with

$$B_{10}^{00} = \begin{bmatrix} D_{0,0} \otimes \mathbf{s}_1 \\ 0 \end{bmatrix},\ B_{10}^{0} = \begin{bmatrix} D_{0,1} \otimes (\mathbf{s}_1\boldsymbol{\beta}_2) \\ 0 \end{bmatrix},\ B_{10}^{1} = \begin{bmatrix} D_{0,0} \otimes (\mathbf{s}_2\boldsymbol{\beta}_2) \\ 0 \end{bmatrix},$$

$$A_0 = \begin{bmatrix} A_0^1 & A_0^0 & & \\ & A_0^1 & A_0^0 & \\ & & \ddots & \ddots \end{bmatrix},\ A_0^1 = \begin{bmatrix} D_{1,0} \otimes S_1 & 0 \\ D_{1,0} \otimes (\mathbf{s}_2\boldsymbol{\beta}_1) & D_{1,0} \otimes S_2 \end{bmatrix},$$

$$A_0^0 = \begin{bmatrix} D_{1,1} \otimes S_1 & 0 \\ D_{1,1} \otimes (\mathbf{s}_2\boldsymbol{\beta}_1) & D_{1,1} \otimes S_2 \end{bmatrix},$$

$$A_2 = \begin{bmatrix} A_2^1 & A_2^0 & & \\ & A_2^1 & A_2^0 & \\ & & \ddots & \ddots \end{bmatrix},\ A_2^1 = \begin{bmatrix} D_{0,0} \otimes (\mathbf{s}_1\boldsymbol{\beta}_1) & 0 \\ 0 & 0 \end{bmatrix},\ A_2^0 = \begin{bmatrix} D_{0,1} \otimes (\mathbf{s}_1\boldsymbol{\beta}_1) & 0 \\ 0 & 0 \end{bmatrix},$$

$$A_1 = \begin{bmatrix} A_1^1 & A_1^0 & & \\ & A_1^1 & A_1^0 & \\ & & \ddots & \ddots \end{bmatrix},\ A_1^0 = \begin{bmatrix} D_{0,1} \otimes S_1 + D_{1,1} \otimes (\mathbf{s}_1\boldsymbol{\beta}_1) & 0 \\ D_{0,1} \otimes (\mathbf{s}_2\boldsymbol{\beta}_1) & D_{0,1} \otimes S_2 \end{bmatrix},$$

$$A_1^1 = \begin{bmatrix} D_{0,0} \otimes S_1 + D_{1,0} \otimes (\mathbf{s}_1\boldsymbol{\beta}_1) & 0 \\ D_{0,0} \otimes (\mathbf{s}_2\boldsymbol{\beta}_1) & D_{0,0} \otimes S_2 \end{bmatrix}.$$

5.3.4.1 Stationary Distribution

If the system is stable we know that there is a relationship between the vector $\boldsymbol{x}$ and the matrix P of the form

$$\boldsymbol{x} = \boldsymbol{x}P, \;\; \boldsymbol{x}\mathbf{1} = 1,$$

and there exists a matrix R for which

$$\boldsymbol{x}_{i+1} = \boldsymbol{x}_i R,$$

where R is the minimal nonnegative solution to the matrix quadratic equation

$$R = A_0 + RA_1 + R^2 A_2.$$

It was shown by Alfa [13] and earlier by Miller [77] for the continuous time case that the matrix R has the following structure

$$R = \begin{bmatrix} R_0 & R_1 & R_2 & R_3 & \cdots \\ & R_0 & R_1 & R_2 & \cdots \\ & & R_0 & R_1 & \cdots \\ & & & R_0 & \ddots \\ & & & & \ddots \end{bmatrix}. \tag{5.114}$$

The arguments leading to this result are as follow. The matrix R is the expected number of visits to the HP state $i+1$ before first returning to the HP state i, given that the process started from state i. By this definition of the matrix R it is straightforward to see how the upper triangular pattern arises, i.e. because the LP number in the system cannot reduce when there is an HP in the system. Since the number of LP arrivals is independent of the state of the number of the HP in the system, the rows of R will be repeating. This structure makes it easier to study the matrix R. However, the $R_i, \; i \geq 0$ also has an additional structure. We have

$$R_i = \begin{bmatrix} R_i(1,1) & 0 \\ R_i(2,1) & R_i(2,2) \end{bmatrix}. \tag{5.115}$$

The matrix R can be computed to a suitable truncation level and then use it to compute the stationary vector after the boundary equation has been solved. We have

$$\boldsymbol{x}_{i+1} = \boldsymbol{x}_i R, \; i \geq 1, \tag{5.116}$$

and specifically we have

$$\boldsymbol{x}_{i+1,j,1} = \sum_{v=0}^{j} [\boldsymbol{x}_{i,j-v,1} R_v(1,1) + \boldsymbol{x}_{i,j-v+1,2} R_v(2,1)], \; i \geq 1, \; j \geq 0, \tag{5.117}$$

and

$$\boldsymbol{x}_{i+1,j+1,2} = \sum_{v=0}^{j} \boldsymbol{x}_{i,j-v+1,2} R_{j-v}(2,2),\ i \geq 1,\ j \geq 0. \tag{5.118}$$

From here on it is straightforward (but laboriously involved) to obtain the stationary distributions and the expected values of the key performance measures. We will skip them in this book and refer the reader to Alfa [13].

5.4 Queues with Multiclass of Packets

In most queueing systems we find that the service disciplines are either FCFS or priority based when we have more than one class of packets. In the last section we dealt with the priority systems of queues already. However sometimes in telecommunications we have multiclass systems with FCFS or sometimes LCFS disciplines. In this section we deal with two classes of multiclass packets – the no priority case and the LCFS. Consider a queueing system with $K < \infty$ classes of packets, with no priority. We assume that the arrival process is governed by the marked Markovian arrival process as defined by He and Neuts [53]. This arrival process is described by $n_t \times n_t$, $K+1$ matrices D_k, $k = 0,1,\cdots,K$, with the elements $(D_k)_{i,j}$ representing the probability of a transition from state i to state j with the arrival of type $k(k = 1,2,\cdots,K)$ and $(D_0)_{i,j}$ represents no arrival. We assume that $0 \leq (D_k)_{i,j} \leq 1$ and $D = \sum_{k=0}^{K} D_k$ is an irreducible stochastic matrix. Let $\boldsymbol{\pi}$ be the stationary vector of the matrix D, then the arrival rate of class k is given by $\lambda_k = \boldsymbol{\pi} D_k \mathbf{1}$. Each class of packet is served according to a PH distribution represented by $(\boldsymbol{\beta}_k, S_k)$ of order $n_s(k)$, $k = 0,1,2,\cdots,K$, with mean service time of class k given as $\mu_k^{-1} = \boldsymbol{\beta}_k (I - S_k)^{-1}\mathbf{1}$. There is a single server attending to these packets. We may have the service in a FCFS or LCFS order. In the next two sections we deal with these two types of service disciplines.

5.4.1 Multiclass systems – with FCFS - the MMAP[K]/PH[K]/1

This class of queues is difficult to analyze using traditional aproaches. Van Houdt and Blondia [100] have extended the idea of age process to analyze this system's waiting time. One may also obtain the queue length information from this waiting time information. In this section we present a summary of the results by van Houdt and Blondia [100].

Define a new matrix S and column vector $\mathbf{s}$, where

$$S = \begin{bmatrix} S_1 & & & \\ & S_2 & & \\ & & \ddots & \\ & & & S_K \end{bmatrix}, \quad \mathbf{s} = \begin{bmatrix} \mathbf{s}_1 \\ \mathbf{s}_2 \\ \vdots \\ \mathbf{s}_K \end{bmatrix}.$$

We now construct a DTMC that represents the age of the leading packet that is in service. We call this the Age process. This process captures how long the packet in service has been in the system up to a point in time. Consider the following state space

$$\Delta = \{(0, J_n^4) \cup (J_n^1, J_n^2, J_n^3, J_n^4),\ J_n^1 \geq 1;$$
$$J_n^2 = 1, 2, \cdots, K; J_n^3 = 1, 2, \cdots, n_s(J_n^2); J_n^4 = 1, 2, \cdots, n_t\}.$$

Here at time n, J_n^1 refers to the age of the packet that is receiving service, J_n^2 the class of packet that is in service, J_n^3 the phase of service of the packet in service, and J_n^4 the arrival phase of the packets. This DTMC has the GI/M/1-type structure with the transition matrix P written as

$$P = \begin{bmatrix} B_0 & C & & & \\ B_1 & A_1 & A_0 & & \\ B_2 & A_2 & A_1 & A_0 & \\ \vdots & \vdots & \vdots & \ddots & \ddots \end{bmatrix}, \tag{5.119}$$

where

$$B_0 = D_0,\ B_j = \mathbf{s} \otimes (D_0)^j,\ j \geq 1,$$
$$C = [(\boldsymbol{\beta}_1 \otimes D_1),\ (\boldsymbol{\beta}_2 \otimes D_2),\ \cdots,\ (\boldsymbol{\beta}_K \otimes D_K)],$$
$$A_0 = S \otimes I_{n_t},\ A_j = \mathbf{s} \otimes ((D_0)^{j-1} C),\ j \geq 1.$$

Provided the system is stable we can easily obtain its stationary distribution using the standard GI/M/1 results presented earlier in Chapter 3. Van Houdt and Blondia [100] did give the conditions under which the system is ergodic and as long as it is ergodic then it is stable. Note that ergodicity in this particular case can only be considered after the state $(0, J_n^4)$ is eliminated from the state space.

5.4.1.1 Stationary distribution and waiting times

In order to proceed with obtaining the stationary distribution we first have to obtain the R matrix which is given as the minimal nonnegative solution to the matrix polynomial equation

$$R = \sum_{j=0}^{\infty} R^j A_j.$$

Let the stationary distribution be $\boldsymbol{x}$ where

$$\boldsymbol{x} = \boldsymbol{x}P,\ \boldsymbol{x}\mathbf{1} = 1,$$

with

$$\boldsymbol{x}_0 = [x_{0,1},\ x_{0,2},\ \cdots,\ x_{0,n_t}],\ \boldsymbol{x}_i = [\boldsymbol{x}_{i,1},\ \boldsymbol{x}_{i,2},\ \cdots,\ \boldsymbol{x}_{i,K}],$$
$$\boldsymbol{x}_{i,j} = [\boldsymbol{x}_{i,j,1},\ \boldsymbol{x}_{i,j,2},\ \cdots,\ \boldsymbol{x}_{i,j,n_s(j)}],\ \boldsymbol{x}_{i,j,\ell} = [x_{i,j,\ell,1},\ x_{i,j,\ell,2},\ \cdots,\ x_{i,j,\ell,n_t}].$$

We follow the standard procedure to obtain the boundary vectors $\boldsymbol{x}_0$ and $\boldsymbol{x}_1$ and then normalize the stationary vector.

Let W_k be the time spent in the system by a class k packet before leaving, i.e. queueing time plus service time. Further define $w_j^k = Pr\{W_k = j\}$, then we have

$$w_j^k = \sum_{v=1}^{n_s(k)} (\lambda_k)^{-1} (\mathbf{t}_k)_v \sum_{u=1}^{n_t} x_{j,k,v,u}, \; j \geq 1, \tag{5.120}$$

where $(\mathbf{t}_k)_v$ is the vth element of the column vector $\mathbf{t}_k$. Note that $w_0^k = 0$ since every packet has to receive a service of at least one unit of time.

In a later paper van Houdt and Blondia [101] presented a method that converts the GI/M/1 structure of this problem to a QBD type. The reader is referred to their paper for more details on this and other related results.

5.4.2 Multiclass systems – with LCFS - the MMAP[K]/PH[K]/1

In a LCFS system with non-geometric service times we need to identify from which phase a service starts again if it is interrupted. Hence we define the matrix Q^k as in the case of the preemptive priority queue earlier, with $Q^k_{\ell_1,\ell_2}$ referring to the probability that given the phase of an ongoing service of a packet of class k is in phase ℓ_1, its future service phase is chosen to start from (after its service resumption) is phase ℓ_2. Note that this Q^k is different from the Q in the previous sections but they both lead to the same concept. However, this current definition makes obtaining some additional performance measures more convenient. The presentation in this section is based on the paper by He and Alfa [52]. It uses the idea and results from tree-structured Markov chains. The continuous time equivalent of this system was discussed in detail by He and Alfa [51] at a limited level. The discrete version in [52] is more general.

Consider a pair (k, j) that is associated with a packet of a type k that is in phase j of service. By being in phase j of service we imply that its service is in phase j if it is currently receiving service or phase j is where its service will start next, if it is currently waiting in the queue. At time n let there be v packets in the system. Associated with this situation let us define

$$q_n = (k_1, j_1)(k_2, j_2)\cdots(k_v, j_v).$$

Here q_n is defined as the queue string consisting of the status of all the packets in the system at the beginning of time n. Let v_n be the phase of the arrival process at time n. It is easy to see that (q_n, v_n) is an irreducible and aperiodic DTMC. Its state space is

$$\Delta = \Omega \times \{1, 2, \cdots, n_t\},$$

where n_t is the phase of arrival and

$$\Omega = \{0\} \cup \{\mathbf{J} : \mathbf{J} = (k_1, j_1)(k_2, j_2) \cdots (k_v, j_v),\ 1 \le k_w \le K,$$
$$1 \le j_w \le n_s(k),\ 1 \le w \le n,\ n \ge 1\}.$$

In fact this DTMC is a tree-structured QBD in the sense that q_n can only increase or decrease by a maximum of 1. We define an addition operation "$+$" here for strings as follows: if we have $\mathbf{J} = (k_1, j_1)(k_2, j_2) \cdots (k_v, j_v)$ then $\mathbf{J} + (k, j) = (k_1, j_1)(k_2, j_2) \cdots (k_v, j_v)(k, j)$.

Let the DTMC be in state $\mathbf{J} + (k, j) \in \Omega$ at time n, i.e. $q_n = \mathbf{J} + (k, j)$, then we have the following transition blocks:

1. When a new packet arrives and there is no service completion, we have

$$A_0((k,j),(k,j')(k_e,j_e))$$
$$\stackrel{\text{def}}{=} Pr\{q_{n+1} = \mathbf{J} + (k,j')(k_e,j_e), v_{t+1} = i' | q_n = \mathbf{J} + (k,j), v_n = i\}$$
$$= (S_k Q^k)_{j,j'} (\beta_{k_e})_{j_e} D_{k_e},\ 1 \le (k,k_e) \le K,\ 1 \le (j,j') \le n_s(k),\ 1 \le j_e \le n_s(k_e), \tag{5.121}$$

2. When a new packet arrives and a service is completed or there is no new packet arrival and there is no service completion, we have

$$A_1((k,j),(k_e,j_e))$$
$$\stackrel{\text{def}}{=} Pr\{q_{n+1} = \mathbf{J} + (k_e,j_e), v_{n+1} = i' | q_n = \mathbf{J} + (k,j), v_n = i\}$$
$$= \begin{cases} (S_k)_{j,j_e} D_0 + ((\mathbf{s}_k)_j \beta_{k_e})_{j_e} D_{k_e}, & k = k_e,\ 1 \le (j,j') \le n_s(k), \\ ((\mathbf{s}_k)_j \beta_{k_e})_{j_e} D_{k_e}, & k \ne k_e,\ 1 \le j \le n_s(k),\ 1 \le j_e \le n_s(k_e) \end{cases}$$

3. When a service is completed and there is no new arrival, we have

$$A_2(k,j) \stackrel{\text{def}}{=} Pr\{q_{n+1} = \mathbf{J}, v_{n+1} = i' | q_n = \mathbf{J} + (k,j), v_n = i\}$$
$$= (\mathbf{s}_k)_j D_0,\ 1 \le k \le K,\ 1 \le j \le n_s(k). \tag{5.122}$$

4. When there is no packet in the system, i.e. $q_n = \mathbf{J} = 0$ we have

$$A_0(0,(k,j)) \stackrel{\text{def}}{=} Pr\{q_{n+1} = (k_e,j_e), v_{n+1} = i' | q_n = 0, v_n = i\}$$
$$= (\beta_{k_e})_{j_e} D_{k_e},\ 1 \le k_e \le K,\ 1 \le j_e \le n_s(k_e). \tag{5.123}$$

$$A_1(0,0) \stackrel{\text{def}}{=} Pr\{q_{n+1} = 0, v_{n+1} = i' | q_t = 0, v_n = i\} = D_0. \tag{5.124}$$

5.4.2.1 Stability conditions

We summarize the conditions under which the system is stable as given by He and Alfa [52]. Analogous to the classical M/G/1 system this current system has a set of matrices $G_{k,j}$, $1 \le k \le K$, $1 \le j \le n_s(k)$, which are minimal non-negative solutions to the matrix equation

$$G_{k,j} = A_2(k,j) + \sum_{k_e=1}^{K} \sum_{j_e=1}^{n_s(k_e)} A_1((k,j),(k_e,j_e))G_{k_e,j_e}$$

$$+ \sum_{j'=1}^{n_s(k)} \sum_{k_e=1}^{K} \sum_{j_e=1}^{n_s(k_e)} A_0((k,j),(k,j')(k_e,j_e))G_{k_e,j_e}G_{k,j'}, \qquad (5.125)$$

and are known to be unique if the DTMC (q_n, v_n) is positive recurrent (see He [50]). Consider the matrices $p((k,j),(k_e,j_e))$ defined as

$$p((k,j),(k_e,j_e)) = \begin{cases} A_1((k,j),(k_e,j_e)) + \sum_{j_e=1}^{n_s(k_e)} A_0((k,j),(k,j')(k_e,j_e)), \; k \neq k_e \\ A_1((k,j),(k,j_e)) + \sum_{j_e=1}^{n_s(k_e)} A_0((k,j),(k,j')(k,j_e)) \\ \quad + A_0((k,j),(k,j_e)(k,j_e))G_{k,j_e} \\ \quad + \sum_{k'=1}^{K} \sum_{j'=1:(k',j')\neq(k,j_e)}^{n_s(k_e)} A_0((k,j),(k,j_e)(k',j'))G_{k',j'}, \\ \qquad k = k_e. \end{cases} \qquad (5.126)$$

Associated with these matrices is another matrix P of the form

$$P = \begin{bmatrix} p((1,1),(1,1)) & p((1,1),(1,2)) & \cdots & p((1,1),(K,n_s(K))) \\ p((1,2),(1,1)) & p((1,2),(1,2)) & \cdots & p((1,2),(K,n_s(K))) \\ \vdots & \vdots & \vdots & \vdots \\ p((K,n_s(K)),(1,1)) & p((K,n_s(K)),(1,2)) & \cdots & p((K,n_s(K)),(K,n_s(K))) \end{bmatrix}. \qquad (5.127)$$

Let $sp(P)$ be the spectral radius of the matrix P, i.e. its largest absolute eigenvalue. Then the Markov chain $\{q_n, v_n\}$ is positive recurrent if and only if $sp(P) < 1$.

5.4.2.2 Performance measures

Here we present an approach for obtaining information about the queue length. First we need to redefine the DTMC in order to make it easier to study the queue length.

We consider the DTMC $\{\tilde{q}_n, v_n, k_n, j_n\}$ where at time n

- $\tilde{q}_n$ is the string consisting of the states of packets waiting in the queue
- v_n is the phase of MMAP[K]
- k_n is the type of packets in service, if any
- j_n is the phase of the service time, if any.

We let the random variable $\tilde{q}_n$ assume the value -1 when there is no packet in the system and assumes the value 0 when there is only one packet in the system. Hence $\tilde{q}_n$ assumes values in the set $\{-1\} \cup \Omega$.

Now we consider the one step block transitions of the DTMC $\{\tilde{q}_n, v_n, k_n, j_n\}$ as follows:

1. $\tilde{q}_n$ going from -1 to -1 and from -1 to 0 as

$$\tilde{A}_1(-1,-1)=D_0,\ \text{ and } \tilde{A}_0(-1,0)=(D_1\otimes\boldsymbol{\beta}_1,\ D_2\otimes\boldsymbol{\beta}_2,\ \cdots,\ D_K\otimes\boldsymbol{\beta}_K),$$

2. $\tilde{q}_t$ going from 0 to -1 as

$$\tilde{A}_2(0,-1)=D_0\otimes\begin{bmatrix}\mathbf{s}_1\\ \mathbf{s}_2\\ \vdots\\ \mathbf{s}_K\end{bmatrix},$$

3. $\tilde{q}_t$ going from $\mathbf{J}+(k,j)$ to $\mathbf{J}+(k,j)(k_e,j_e)$, or to $\mathbf{J}+(k,j)$, or to $\mathbf{J}$, $\mathbf{J}\in\Omega$ as

$$\tilde{A}_0(k_e,j_e)=\sum_{w=1}^{K}D_w\otimes\left[\begin{pmatrix}0\\ \vdots\\ (S_{k_e}Q_{k_e})_{j_e}\boldsymbol{\beta}_w\\ 0\\ \vdots\\ 0\end{pmatrix}(0,\ \cdots,\ 0,\ I_{n_s(w)},\ 0\cdots,\ 0)\right];$$

$$\tilde{A}_1(k,j)=D_0\otimes\begin{pmatrix}S_1&&&\\ &S_2&&\\ &&\ddots&\\ &&&S_K\end{pmatrix}$$

$$+\sum_{w=1}^{K}D_w\otimes\left[\begin{pmatrix}\mathbf{s}_1\boldsymbol{\beta}_w\\ \vdots\\ \mathbf{s}_K\boldsymbol{\beta}_w\end{pmatrix}(0,\ \cdots,\ 0,\ I_{n_s(w)},\ 0\cdots,\ 0)\right];$$

$$\tilde{A}_2(k,j)=D_0\otimes\left[\begin{pmatrix}\mathbf{s}_1\mathbf{e}_j\\ \vdots\\ \mathbf{s}_K\mathbf{e}_j\end{pmatrix}(0,\ \cdots,\ 0,\ I_{n_s(w)},\ 0\cdots,\ 0)\right],$$

where $(S_{k_e}Q_{k_e})_{j_e}$ is the column j_e of the matrix $S_{k_e}Q_{k_e}$. One notices that matrix $\tilde{A}_1(k,j)$ is independent of (k,j).

Let

$$\boldsymbol{x}(\mathbf{J},i,k,j))=Pr\{(\tilde{q}_n,v_n,k_n,j_n)=(\mathbf{J},i,k,j))\}|_{n\to\infty};$$

and

$$\boldsymbol{x}(-1,i)=Pr\{\{(\tilde{q}_n,v_n))=(-1,i)\}|_{n\to\infty}.$$

Further define

1.

$$\boldsymbol{x}(\mathbf{J},i,k)=[\boldsymbol{x}(\mathbf{J},i,k,1),\ \boldsymbol{x}(\mathbf{J},i,k,2),\ \cdots,\ \boldsymbol{x}(\mathbf{J},i,k,n_s(k))],$$

2.

$$\boldsymbol{x}(\mathbf{J},i) = [\boldsymbol{x}(\mathbf{J},i,1),\ \boldsymbol{x}(\mathbf{J},i,2),\ \cdots,\ \boldsymbol{x}(\mathbf{J},i,K)],$$

3.

$$\boldsymbol{x}(\mathbf{J}) = [\boldsymbol{x}(\mathbf{J},1),\ \boldsymbol{x}(\mathbf{J},2),\ \cdots,\ \boldsymbol{x}(\mathbf{J},n)],\ \mathbf{J} \neq -1;$$

4.

$$\boldsymbol{x}(-1) = [\boldsymbol{x}(-1,1),\ \boldsymbol{x}(-1,2),\ \cdots,\ \boldsymbol{x}(-1,n_t)].$$

Let $R = \sum_{k=1}^{K} \sum_{j=1}^{n_s(k)} R_{k,j}$ where the matrix sequence $\{R_{k,j},\ 1 \leq j \leq n_s(k),\ 1 \leq k \leq K\}$ is the minimal non-negative solutions to the matrix equations

$$R_{k,j} = \tilde{A}_1(k,j) + R_{k,j}\tilde{A}_1 + R_{k,j} \sum_{k=1}^{K} \sum_{j=1}^{n_s(k)} R_{k_e,j_e} \tilde{A}_2(k_e,j_e). \tag{5.128}$$

The computations of $\{R_{k,j},\ 1 \leq j \leq n_s(k),\ 1 \leq k \leq K\}$ can be carried out using the algorithm presented in Yeung and Alfa [103] or the one in Yeung and Sengupta [104].

The probability distribution of the queue length and the associated auxilliary variables can be obtained from

1.
$$\boldsymbol{x}(\mathbf{J}+(k,j)) = \boldsymbol{x}(\mathbf{J})R_{k,j},\ \ \mathbf{J} \in \Omega,\ 1 \leq k \leq K,\ 1 \leq j \leq n_s(k); \tag{5.129}$$

2.

$$\boldsymbol{x}(0) = \boldsymbol{x}(-1)\tilde{A}_0(-1,0) + \boldsymbol{x}(0)\tilde{A}_1(0,0) + \sum_{k=1}^{K} \sum_{j=1}^{n_s(k)} \boldsymbol{x}(0)R_{k,j}\tilde{A}_2(k,j), \tag{5.130}$$

3.
$$\boldsymbol{x}(-1) = \boldsymbol{x}(-1)\tilde{A}_1(-1,-1) + \boldsymbol{x}(0)\tilde{A}_2(0,-1); \tag{5.131}$$

4.
$$\boldsymbol{x}(-1)\mathbf{1} + \boldsymbol{x}(0)(I-R)^{-1}\mathbf{1} = 1, \tag{5.132}$$

References

1. N. Akar and K. Sohraby, An invariant subspace approach in M/G/1 and GI/M/1 type Markov chains *Stochastic Models*, 13, 381-416, 1997.
2. A. S. Alfa and B. H. Margolius, Two classes of time-inhomogeneous Markov chains: Analysis of the periodic case *Annals of Oper. Res.*, 160: 121-137, 2008.
3. A. S. Alfa and J. Xue, Efficient computations for the discrete GI/G/1 system *INFORMS Journal on Computing*, 19, 480-484, 2007.
4. A. S. Alfa, B. Liu and Q-M. He, Discrete-time analysis of MAP/PH/1 multiclass general preemptive priority queue *Naval Research Logistics*, 50, 662-682, 2003.
5. A. S. Alfa, Discrete time queues and matrix-analytic methods *Sociedad de Estadistíca e Investigaciòn Operativa TOP*, 10, 147-210, 2002.
6. A.S. Alfa, Markov chain representations of discrete distributions applied to queueing models. *Computers and Operations Research*, 31:2365-2385, 2004.
7. A.S. Alfa and S. Chakravarthy, A discrete queue with Markovian arrival processes and phase type primary and secondary services. *Stochastic Models*, 10:437–451, 1994.
8. A.S. Alfa, K.L. Dolhun, and S. Chakravarthy, A discrete single server queue with Markovian arrivals and phase type group service. *Journal of Applied Mathematics and Stochastic Analysis*, 8:151–176, 1995.
9. A.S. Alfa and I. Frigui, Discrete NT-policy Single Server Queue with Markovian Arrival Process and Phase Type Service. *European Journal of Operational Res.*, 88, 599-613, 1996.
10. A.S. Alfa and M.F. Neuts, Modelling vehicular traffic using the discrete time Markovian arrival process. *Transportation Science*, 29:109–117, 1995.
11. A. S. Alfa, A discrete MAP/PH/1 queue with vacations and exhaustive time-limited service *Operations Research Letters*, 18, 31-40, 1995
12. A. S. Alfa, A discrete MAP/PH/1 vacation queue with gated time-limited service *Queueing Systems - Theory and Applications*, 29, 35-54, 1998
13. A. S. Alfa, Matrix-geometric solution of discrete time MAP/PH/1 priority queue *Naval Research Logistics*, 45, 23-50, 1998
14. A. S. Alfa and W. Li, Matrix-geometric analysis of the discrete GI/G/1 system *Stochastic Models*, 17, 541-554, 2001.
15. A. S. Alfa, B. Sengupta, T. Takine and J. Xue, A new algorithm for computing the rate matrix of GI/M/1 type Markov chains Proceedings of the 4th International Conference on matrix Analytic Methods, Adelaide, Australia, 1-16, 2002.
16. A. S. Alfa, B. Sengupta and T. Takine, The use of non-linear programming in matrix analytic methods Stochastic Models, 14, 351-367, 1998.
17. A. S. Alfa, An alternative approach for analyzing finite buffer queues in discrete time *Performance Evaluation*,53:75-92, 2003.
18. A. S. Alfa, Algorithmic analysis of the BMAP/D/k system in discrete time, *Adv. Appl. Probab.*, 35: 1131-1152, 2003.
19. J.R. Artalejo and O. Hernández-Lerma, Performance analysis and optimal control of the Geo/Geo/c queue Performance Evaluation, 52, 15-39, 2003.
20. S. Asmussen and O. Nerman, Fitting phase-type distributions via the EM algorithm. In *Symposium i Anvendt Statiskik, K. Vest Nielsen (ed.)*, pages 335–346, Copenhagen, January 1991.
21. N. U. Bhat, *Elements of Applied Stochastic Proceses*, Wiley, New York, 1972.
22. D. Bini and B. Meini, On the solution of a nonlinear matrix equation arising in queueing problems, *SIAM Jour. Matrix Anal. Appl.*, 17: 906-926.
23. C. Blondia, A discrete-time batch Markovian arrival process as B-ISDN traffic model. *Belgian Journal of Operations Research, Statistics and Computer Science*, 32:3–23, 1992.
24. A. Bobbio and A. Cumani, ML estimation of the parameters of a PH distribution in triangular canonical form. *Computer Performance Evaluation, G. Balbo an G. Serazzi (ed.), Elsevier Science Pub. B.V.*, pages 33–46, 1992.

25. A. Bobbio and M. Telek, Parameter estimation of phase type distributions. In *20th European Meeting of Statisticians*, pages 335–346,1992. Bath, Uk, September 1992.
26. A. Bobbio, A. Hovarth, M. Scarpa and M. Telek, Acyclic discrete phase type distributions: properties and a parameter estimation algorithm. In *Performance Evaluation*,54, 1 -32, 2004.
27. L. Breuer and A. S. Alfa, An EM algorithm for platoon arrival processes in discrete time, *Operations Res. Lett.*, 33: 535-543, 2005.
28. H. Bruneel and B. G. Kim, *Discrete-time Models for Communication systems Including ATM* Kluwer Academic Pub., Boston, 1993.
29. S. Chakravarthy and A.S. Alfa, A multiserver queue with Markovian arrival process and group services with thresholds. *Naval Research Logistics*, 40:811–827, 1993.
30. S. Chakravarthy and A.S. Alfa, A finite capacity queue with Markovian arrivals and two servers with group services. *Journal of Applied Mathematics and Stochastic Analysis*, 7:161–178, 1994.
31. S. Chakravarthy and M.F. Neuts, Algorithms for the design of finite-capacity service units. *Naval Research Logistics*, 36:147–165, 1989.
32. M. L. Chaudhry, On numerical computations of some discrete-time queues. *Computational Probability*,Kluwer's International Series, W. K. Grassmann (Ed.), 365-408, 2000
33. C. D. Crommelin, Delay probability formulae when the holding times are constant. *Post Off. Electr. Eng. J.*, 25:41-50, 1932.
34. H. Daduna, *Queueing Networks with Discrete Time Scale: Explicit Expressions for the Steady State Behaviour of Discrete Time Stochastic Networks*, Springer Lecture Notes in Computer Science, 2001.
35. S.C. Dafermos and M.F. Neuts, A single server queue in discrete time. *Cahiers Centre tudes Rech. Opr.*, 13:23–40, 1971.
36. B.T. Doshi, A Note on Stochastic Decomposition in a GI/G/1 Queue with Vacations or Set-up Times. *J. Appl. Prob.*, 22:419–428, 1985.
37. B.T. Doshi, Queueing Systems with Vacations - A Survey. *Queueing Systems*, 1:29–66, 1986.
38. B.T. Doshi, *Stochastic Analysis of Computer and Communications Systems*, A chapter in Single Server Queues with Vacations. H. Takagi edition, 1990.
39. A. K. Erlang, The theory of probabilities and telephone conversations, *Nyt Tidsskrift Mat. B* 20, 33-39.
40. A. K. Erlang, Solution of some problems in the theory of probabilities of significance in automatic telephone exchanges, *Electrotejnikeren (Danish), 13, 5-13. (English Translation, P. O. Elec. Eng. J. 10, 189-197 (1917-1918)*, 13, 5-13, 1917.
41. I. Frigui and A.S. Alfa, Analysis and Computation of the Busy Period for the Discrete MAP/PH/1 Queue with Vacations. *manuscript*, 1995.
42. S. W. Fuhrman and R. B. Cooper, Stochastic decomposition in the M/G/1 queue with generalized vacations *Oper. Res.*, 33:1117-1129, 1985.
43. H. R. Gail, S. L. Hantler and B. A. Taylor, Non-skip-free M/G/1 and GI/M/1 type Markov chains Adv. Appl. Prob., 29, 733-758, 1997.
44. H.P. Galliher and R.C. Wheeler, Nonstationary queuing probabilities for landing congestion of aircraft. *Operations Research*, 6:264–275, 1958.
45. D. P. Gaver, P. A. Jacobs abd G. Latouche, Finite birth-and-death models in randomly changing environments, *Adv. Appl. Probab.*, 16:715-731, 1984.
46. W.K. Grassmann, Taksar and D.P. Heyman, Regenerative analysis and steady state distributions for Markov chains. *Operations Research*, 33:1107-1117.
47. W.K. Grassmann and D.P. Heyman, Equilibrium distribution of block-structured Markov chains with repeating rows. *J. Appl. Prob.*, 27:557–576, 1990.
48. W.K. Grassmann and D.P. Heyman, Computation of steady-state probabilities for infinite-state Markov chains with repeating rows. *ORSA J. on Computing*, 5:292–303, 1993.
49. D. Gross, J. F. Shortle, J. M. Thompson and C.M. Harris, *Fundamentals of Queueing Theory*. John Wiley and Sons, New York, 2nd edition, 2008.
50. Q-M. He, Classification of Markov processes of matrix M/G/1 type with a tree structure and its applications to the MAP[K]/PH[K]/1 queue Internal Report, Dalhousie University, 1998.

51. Q-M. He and A. S. Alfa, The MMAP[K]/PH[K]/1 queue with a last-come-first-served preemptive service discipline Queueing Systems - Theory and Applications, 28, 269-291, 1998.
52. Q-M. He and A. S. Alfa, The discrete time MMAP[K]/PH[K]/1/LCFS-GPR queue and its variants *Advances in algorithmic Methods for Stochastic Models,* Notable Pub. Inc, G. latouche and P. Taylor (Ed), 167-190, 2000.
53. Q-M. He and M. F. Neuts, Markov arrival process with marked transitions Stochastic Processes and Applications, 74, 37-52, 1998.
54. D. Heimann and M.F. Neuts, The Single Server Queue in Discrete Time Numerical Analysis IV. *Naval Research Logistics Quarterly*, 20:753–766, 1973.
55. D.P. Heyman, Accurate computation of the fundamental matrix of a Markov chain. *SIAM Jour. Matrix Anl. Appl.*, 16:151-159, 1995.
56. R. Howard, *Dynamic Programming and Markov Processes*. MIT Press, Cambridge, 1960.
57. J.J. Hunter, *Mathematical Techniques of Applied Probability - Volume 2, Discrete Time Model: Techniques and Applications*. Academic Press, New York, 1983.
58. J.R. Jackson, Network or Waiting Lines. *Operations Research*, 5:518–521, 1957.
59. J.R. Jackson, Jobshop-like Queueing Systems. *Management Science*, 10:131–142, 1963.
60. M. Johnson and M. Taaffe, Matching moments to phase distributions: mixtures of erlang distributions of common order. *Stochastic Models*, 5, 1989.
61. D. G. Kendall, Some problems in the theory of queues. *J. Roy. Statist. Soc. Ser. B*, 13: 151-185, 1951.
62. D. G. Kendall, Stochastic processes occurring in the theory of queues and their analysis by the method of imbedded Markov chains. *Ann. Math. Statist.*, 24: 338-354, 1953.
63. L. Kleinrock, *Queueing Systems, Volume 1: Theory*. John Wiley and Sons, New York, 1975.
64. M.F. Klimko and M.F. Neuts, The single server queue in discrete time - numerical analysis, II. *Naval Res. Logist. Quart.*, 20:304–319, 1973.
65. G. Latouche and V. Ramaswami, A logarithmic reduction algorithm for quasi-birth-and-death processes. *J. Appli. Probab.*, 30:650-674, 1993.
66. G. Latouche and V. Ramaswami, The PH/PH/1 queue at epochs of queue size change *Queueing Systems, Theory and Appl.*, 25: 97-114, 1997
67. G. Latouche and V. Ramaswami, *Introduction to matrix Analytic Methods in Stochastic Modeling* ASA-SIAM Series on Statistics and Applied Probability, Philadelphia, Pennsylvania, 1999.
68. Y. Levy and U. Yechiali, Utilization of Idle Time in an M/G/1 Queueing System. *Management Science*, 22:202–211, 1975.
69. L. R. Lipsky, *Queueing Theory: A Linear Algebraic Approach* Macmillan Publishing Company, 1992.
70. Q. Liu, A. S. Alfa amd J. Xue, Analysis of the discrete-time GI/G/1/K using remaining time approach *J. Appl. Math and Informatics*, vol. 28, No. 1-2, 153-162, 2010.
71. D. Liu and M.F. Neuts, A queuing model for an ATM rate control scheme. *Telecommunication Systems*, 1995.
72. D.M. Lucantoni, New results on single server queue with a batch Markovian arrival process. *Stochastic Models*, 7:1–46, 1991.
73. D.M. Lucantoni, K.S. Meier-Hellstern, and M.F. Neuts, A single server queue with server vacations and a class of non-renewal arrival processes. *Advances in App. Prob.*, 22:676–705, 1990.
74. D.M. Lucantoni and M.F. Neuts, Some steady-state distributions of the MAP/SM/1 queue. *Stochastic Models*, 10, 1994.
75. A. A. Markov, Extension of limit theorems of probability theory to sum of variables connected in a chain, *The Notes of the Imperial Academy of Sciences of St. Petersburg VIII Series, Physio-Mathematical College*, XXII, no. 9, December 5, 1907.
76. B. Meini, Solving M/G/1 type Markov chains: Recent advances and applications *Stochastic Models*, 14, 479-496, 1998.
77. D. G. Miller, Computation of steady-state probabilities for the M/M/1 priority queues, *Operations Research*, 29: 945-958, 1981.

78. D.L. Minh, The discrete-time single-server queue with time-inhomogeneous compound Poisson input. *Journal of Applied Probability*, 15:590–601, 1978.
79. M.F. Neuts, A versatile Markovian point process. *Journ. Appl. Prob.*, 16:764–779, 1979.
80. M.F. Neuts, *Matrix-geometric solutions in stochastic models - An algorithmic approach.* The Johns Hopkins University Press, Baltimore, 1981.
81. M.F. Neuts, *Structured Stochastic Matrices of M/G/1 Type and Their Applications.* Marcel Decker Inc., New York, 1989.
82. M.F. Neuts, Models based on the Markovian arrival process. *IEICE Trans. Commun.*, E75-B:1255–1265, 1992.
83. M.F. Neuts and M.F. Klimko, The single server queue in discrete time - numerical analysis, III. *Naval Res. Logist. Quart.*, 20:297–304, 1973.
84. M. F. Neuts, Probability distributions of phase type. In *Liber Amicorum Prof. Emeritus H. Florin*, 173-206, University of Louvain, Belgium, 1975.
85. M.F. Neuts, *Probability* Allyn and Bacon, Biston, 1973.
86. Park, D., Perros H. and H. Yamashita, Approximate analysis of discrete-time tandem queuing networks with bursty and correlated input traffic and customer loss. *Oper. Res. Letters*, 15:95–104, 1994.
87. H.G. Perros and T. Altiok (ed.), Queueing Networks with Blocking. 1989.
88. Md. M. Rahman and A. S. Alfa, Computational procedures for a class of GI/D/k system in discrete time, J. Probab. & Statis., vol. 2009, Article ID: 716364, 18 pages, 2009, doi:10.1155/2009/716364, 2009.
89. V. Ramaswami, A stable recursion for the steady state vector in Markov chains of M/G/1 type. *Stochastic Models*, 4:183–188, 1988.
90. V. Ramaswami and G. Latouche, A General Class of Markov Processes with Explicit Matrix-Geometric Solutions. *OR Spektrum*, 8:209–218, 1986.
91. D. Shi, J. Guo and L. Liu, SPH-distributions and the rectangle-iterative algorithm *Matrix-analytic Methods in Stochastic Models*, Chakravarthy, S. and Alfa, A. S. (Ed.), Marcel Dekker, 207-224, 1996.
92. D. Shi and D. Liu, Markovian models for non-negative random variables *Advances in Matrix-analytic Methods in Stochastic Models*, Alfa, A. S. and Chakravarthy, S. (Ed.), Notable Publications, 403-427, 1998.
93. W.J. Stewart, *Introduction to the numerical solution of Markov chains.* Princeton University Press, Princeton, N.J., 1994.
94. H. Takagi, *Queuing Analysis, A Foundation of Performance Analysis - Volume 3*, chapter Discrete-Time Systems. North Holland, Amsterdam, 1993.
95. T. Takine, B. Sengupta and R. W. Yeung, A generalization of the matrix M/G/1 paradigm for Markov chains with a tree structure, *Stochastic Models*, 11: 411-421, 1995.
96. M. Telek, The minimal coefficient of variation of discrete phase type distributions 3rd Conf. on MAM, 391-400, 2000.
97. N. Tian and Z. G. Zhang, The discrete-time GI/Geo/1 queue with multiple vacations *Queueing Systems*, 40, 283-294, 2002.
98. H. C. Tijms, *Stochastic Models, an Algorithmic Approach* Wiley, New York, 1994.
99. R. L. Tweedie, Operator-geometric stationary distributions for Markov chains, with applications to queueing models. *Adv. Appl. Proab.*, 14:368-391, 1982.
100. B. Van Houdt and C. Blondia, The delay distribution of a type k customer in a first-come-first-served MMAP[K]/PH[K]/1 queue *Journal of Applied Probability*, 39, 213-223, 2002.
101. B. Van Houdt and C. Blondia, The waiting time distribution of a type k customer in a discrete-time MMAP[K]/PH[K]/c (c=1,2) queue using QBDs *Stochastic Models*, 20, 1, 55-69 2004.
102. J. Ye and S.-Q. Li, Folding algorithm: A computational method for finite QBD processes with level dependent transitions. *IEEE Trans. Commun.*, 42: 625-639, 1994.
103. R. W. Yeung and A. S. Alfa, The quasi-birth-and-death type Markov chain with a tree structure *Stochastic Models*, 15, 639-659, 1999.
104. R. W. Yeung and B. Sengupta, Matrix product-form solutions for Markov chains with a tree structure, *Adv. Appl. Probab.*, 26: 965-987, 1994.

105. Z. G. Zhang and N. Tian, Discrete-time Geo/GI/1 queue with multiple adaptive vacations *Queueing Systems*, 38, 419-429, 2001.
106. R. W. Wolff, *Stochastic Modelling and the Theory of Queues*. Prentice hall, Engelwood Cliffs, New Jersey, 1989.
107. Y.Q. Zhao, W. Li, and W. J. Braun, Infinite block-structured transition matrices and their properties. *Adv. Appl. Prob.*, 30:365-384, 1998.
108. Y.Q. Zhao, W. Li, and A.S. Alfa, Duality results for block-structured transition matrices. *J. Appl. Prob.*, 36:1045-1057, 1999.

Index

A.S. Alfa, *Queueing Theory for Telecommunications*,
DOI 10.1007/978-1-4419-7314-6, © Springer Science+Business Media, LLC 2010